Nuclear Power
in Perspective

NUCLEAR POWER IN PERSPECTIVE

Eric Addinall and Henry Ellington

Kogan Page, London
Nichols Publishing Company, New York

Dedication
This book is dedicated to our children – Stephen, Kenneth, Mary Jane and Pamela – whose lives will almost certainly be cleaner, healthier and more prosperous as a result of nuclear power.

First published in Great Britain in 1982 by
Kogan Page Ltd, 120 Pentonville Road, London N1 9JN

British Library Cataloguing in Publication Data
Addinall, Eric
Nuclear power in perspective.
1. Atomic power – Social aspects
I. Title II. Ellington, Henry
306'.4 TK9153

ISBN 0-85038-510-5

First published in the United States of America in 1982 by
Nichols Publishing Company, PO Box 96, New York, NY 10024

Library of Congress Cataloging in Publication Data
Addinall, Eric.
Nuclear power in perspective.

1. Atomic power. I. Ellington, Henry. II. Title.
TK9146.A32 333.79'24 81-11099
ISBN 0-89398-110-3 AACR2

Printed in Great Britain
by the Anchor Press Ltd, and bound
by William Brendon & Son,
both of Tiptree, Essex

Contents

Acknowledgements

The authors would like to acknowledge the assistance they received from the following in producing this book:

- Norman H Langton, Professor of Physics at Robert Gordon's Institute of Technology, for continuous advice, support and encouragement and for constructive criticism during the preparation of the manuscript;
- the United Kingdom Atomic Energy Authority and British Nuclear Fuels Ltd, for providing the bulk of the photographs, acting as an invaluable source of information, and arranging for each section of the manuscript to be criticized by members of their staff; particular thanks are due to John Bird, Peter McLaren and Eric Willmott;
- Margaret Geddes, for typing the manuscript.

The authors and publisher would like to thank the following for permission to reproduce copyright material:

- The British Embassy, Washington, DC, for Figure 11.2(a);
- British Nuclear Fuels Ltd, for Figures 3.2, 3.3, 3.6, 8.1 and 9.5;
- Hydro-Quebec International, Quebec, Canada, for Figure 2.8(b);
- JET Joint Undertaking, for Figures 2.11(a) and 2.11(b);
- Northeast Utilities, Connecticut, USA, for Figure 2.7(b);
- The North of Scotland Hydro-Electric Board, for Figure 7.9;
- Shell UK Exploration and Production, for Figure 7.2;
- The United Kingdom Atomic Energy Authority, for Figures 2.5, 2.6(a), 2.6(b), 2.7(a), 2.8(a), 2.10, 3.4, 4.4, 4.5, 7.3, 7.5, 7.6, 7.8, 11.2(b), 11.3, 11.6, 12.1 and 12.2;
- The United States Department of Energy for Figures 11.4 and 11.5.

Introduction

During most of our working lives, we have been heavily involved in energy-related work – both through our teaching and through our research and consultancy. The latter has involved us in working with several organizations which have a vested interest in keeping society supplied with the energy it needs, and has put us in the privileged position of being able to develop an overview of the 'energy scene' with the benefit of inside knowledge.

In the course of our work, we have become increasingly aware of the fact that one particular energy source – nuclear power – consistently gets a 'bad press'. Society accepts that we need coal, oil and natural gas, and quite cheerfully puts up with the loss of life, pollution and reduction of amenity that their provision inevitably entails – just as it accepts the fearful death toll exacted by the motor car. Similarly, most people accept that we need to develop the various 'alternative' sources of energy, despite the fact that many of the proposed schemes are open to serious criticism on environmental and amenity grounds. When we come to nuclear power, however – arguably the cleanest, safest and cheapest source of energy that man has yet developed – the story is different, and we seem to hear nothing but objections and protests. Admittedly, the most vociferous of these protests invariably comes from banner-waving minority groups whose knowledge of the issues seldom matches the fervour with which they express their views, but large numbers of ordinary citizens are also genuinely concerned about the alleged dangers associated with nuclear power. Indeed, such is the current strength of anti-nuclear feeling (particularly in the USA) that the whole future of nuclear power must be in doubt.

As objective observers of the energy field, we view this development with some alarm, because our work has led us to the firm conclusion that the Western world is soon going to need all the energy it can produce – including nuclear power. At the moment, it is certainly true that we have rather more energy than we need – largely as a result of the traumatic events that followed the Yom Kippur War and the current world recession. This situation will not last, however, and, by the end of the century, we seem likely to be faced with a severe energy shortfall unless we take active steps to replace our dwindling supplies of fossil fuels with new sources of energy. Not to make the maximum possible use of nuclear power in these circumstances would, in our opinion, be sheer folly.

As we see it, there are two main reasons why people tend to be afraid of nuclear power. Firstly, they inevitably associate the word 'nuclear' with nuclear weapons, so that the perfectly natural fear that we all have of nuclear war tends to be extended (albeit subconsciously in most cases) to include the peaceful applications of nuclear energy. Secondly, the general public have not so far been given a chance to carry out an objective appraisal of the nuclear power issue, having been subjected to a well-organized, systematic campaign of anti-nuclear propaganda that has not, unfortunately, been balanced by a similarly powerful statement of the pro-nuclear case. Indeed, governments have, in the main, been extremely reluctant to make information about nuclear power generally available. We believe that if people were made aware of *all the facts* concerning the nature of nuclear power and the role it could play in helping to overcome the impending energy crisis, the current anti-nuclear mood would largely disappear. We have written this book in the hope that it will make some contribution to bringing about such a change of mood.

Numerous books on nuclear power have been published in recent years, but, with very few exceptions, they have either been highly specialized or have presented only one side (or one aspect) of the nuclear argument. As far as we are aware, no one has so far written a book that describes, in simple, non-technical language, exactly what nuclear power is, and gives an objective survey of the main arguments that are commonly made for and against nuclear power. This book attempts to do both these things, and should give the informed layman the information that he needs in order to make up his own mind on the nuclear power issue.

The book consists of four main parts, plus an appendix. *Part I* is a simple introduction to nuclear power, and is designed to provide the non-technical reader with sufficient background knowledge of the field to enable him to follow the arguments that are presented later. Chapter 1 outlines the physical principles on which nuclear power is based, describing the structure of the atom and explaining how it is possible to release energy from its nucleus. Chapter 2 then describes how this energy is harnessed in nuclear reactors, following a general description of the principles of reactor design with a survey of the main types that are currently in use or being developed; the chapter ends by taking a brief look at the prospects of exploiting fusion power. Next, Chapter 3 deals with the general topic of nuclear fuel, describing how such fuel is manufacturerd and discussing the various possible fuel cycles. Finally, Chapter 4 explains what radioactivity is, and describes how ionizing radiation can affect living tissue.

Part II is a broad survey of the general energy field, and asks the basic question: 'do we really need nuclear power?' It uses the USA and Britain as case studies. Chapter 5 sets the scene by looking at how we have used energy during the last 30 years or so, and Chapter 6 then tries to forecast how this usage pattern is likely to change over the next 20 years by examining the implications of two extreme scenarios – 'historic growth' and 'zero growth'. Finally, Chapter 7 tries to determine whether we are likely to be able to meet our future energy requirements by relying solely on non-nuclear sources, or whether we will *have* to make extensive use of nuclear power.

Part III then outlines the nuclear options that are available to the Western world, again using the USA and Britain as case studies. Chapter 8 sets the scene by describing how nuclear power has developed to date, and Chapter 9 outlines the contributions that thermal and fast reactors could make towards meeting our future energy needs, presenting the case for a balanced, 'mixed' nuclear power programme.

Finally, *Part IV* deals with what is to most people the key issue, namely, 'is nuclear power socially and environmentally acceptable?' Chapter 10 examines the question of whether or not workers in the nuclear power industry are subjected to unacceptably high accident and/or health risks. Next, Chapter 11 looks at the alleged danger to the general public, dealing in turn with the actual operation of nuclear power stations, the transportation of nuclear materials, the reprocessing of nuclear fuel and the disposal of nuclear waste; the chapter concludes by discussing the alleged genetic hazard represented by the nuclear power industry. Finally, Chapter 12 discusses the question of whether nuclear power constitutes a threat to democratic freedom and examines the so-called proliferation issue.

In order to provide readers with a readily accessible source of reference material, a comprehensive glossary of terms used in the nuclear power industry is included as an appendix to our book. This is again written in non-technical language. We have also included a short bibliography for the benefit of those readers who wish to study any of the topics covered in greater depth.

PART ONE

The Nature of Nuclear Power

In order to follow the finer points of 'the nuclear debate', it is first necessary to understand the nature of nuclear power. Part I of our book is intended to provide the basis of such an understanding.

If you are already familiar with such topics as the nature of the atomic nucleus, fission and fusion, the workings of nuclear reactors, the nuclear fuel cycle, and the nature and biological effects of radioactivity, you may pass to Part II without disadvantage, since you can always refer back if the need arises.

If, however, you are unfamiliar with the language and technology of nuclear power, we hope that Part I will provide the introduction and background that will enable you to follow the (often conflicting) arguments put forward later in the book, and, we hope, put you in a position to make up your own mind on the various issues examined.

CHAPTER ONE

The Atomic Nucleus – a Potential Source of Energy

In this chapter, we shall look at the building blocks from which all matter is made – *atoms* – and at their detailed structure, so that we can learn how they can be persuaded to give up part of the energy in the nuclei to generate electricity.

The atomic nature of matter

Although the Greeks are probably remembered more for their contributions to art, literature, mathematics and philosophy than for their writings on the natural sciences, it is to Democritus (460-370 BC) that we owe the germ of the atomic theory of matter. He postulated that the universe consists of empty space and an almost infinite number of indivisible and invisible particles (atoms) which differ from each other in form, position and arrangement. The Greeks argued that there were as many different kinds of atoms as there were fundamentally different substances (in the fourth century BC, these amounted to fire, air, water and earth). Unfortunately, no evidence could be found for the existence of atoms, and 'the atomic hypothesis' remained nothing more than an abstract metaphysical idea for over two thousand years.

John Dalton (1766-1844), a chemist, was responsible for reviving the atomic theory of matter, and is now recognized as the pioneer of the modern concept of the atom. Like the Greeks, he argued that everything was made up of atoms which combined to make all the substances we know, but his atoms were of hydrogen, oxygen, carbon and so on, and not fire, water, etc.

We now know that all matter, whether solid, liquid or gas, is made up of tiny particles (atoms) which combine to form larger (but still tiny) particles called *molecules*. These atoms are extremely small – typically measuring less than one ten-millionth of a centimetre in diameter. Indeed, if people were as small as atoms, the entire population of the world could stand on a single full stop on this page with plenty of room to spare.

Substances which consist of only one type of atom are known as *elements*, and cannot be split up into simpler substances by chemical means. Uranium and plutonium (of which we shall be hearing much more) are examples of elements, as are iron, hydrogen and oxygen. In fact, there are over one hundred different elements, but only 92 of these occur naturally.

All other substances consist of combinations or *compounds* of these elements. Water, for example, is a compound with molecules consisting of two atoms of hydrogen and one of oxygen.

The structure of the atom

Until around 1900, the atom was thought of as a hard, impenetrable billiard ball-like body which was electrically neutral (ie had no net positive or negative charge). However, the discovery of radioactivity in 1896 (see Chapter 4) and the work of Lord Rutherford in 1909 changed this picture dramatically.

It was discovered that the atom had an internal structure, and that atoms of one kind could be changed into other kinds of atoms by altering this structure. (Some atoms even changed their internal structure without outside help – a phenomenon known as *natural radioactivity*.) Although the atom was known to be electrically neutral, it was found that electrons could be removed from it, leaving it positively charged. The picture was made even more perplexing by the discovery that practically the entire volume of an atom was made up of empty space.

The results of many subsequent experiments showed that the internal structure of the atom is well defined. Practically all of its mass is contained in a central, positively charged *nucleus*, a tiny body less than one ten-thousandth of the diameter of the atom as a whole. This is surrounded by a diffuse cloud of orbiting *electrons* – negatively charged particles that are bound to the nucleus by electrical forces. Figure 1.1 shows a schematic representation of this structure.

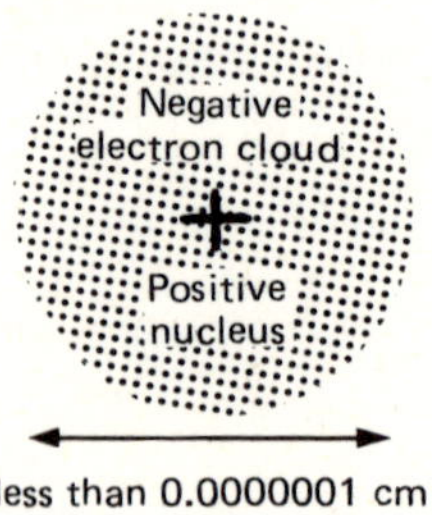

Figure 1.1 *The structure of an atom*

The structure of the nucleus

Later experiments showed that the nucleus also had an internal structure. The nature of this structure took a long time to determine, however, largely because of difficulties in reconciling the measured masses of nuclei with their observed charges. Indeed, it was not until 1932, when Chadwick discovered a new particle which he named the *neutron* (because it carried no electrical charge), that a satisfactory model of the nucleus was developed.

The atomic nucleus is now known to have a well-defined internal structure, being composed of positively charged particles called *protons* and uncharged neutrons. While both the protons and the neutrons contribute to the *mass* of the nucleus, the nuclear *charge* comes only from the protons. The number of protons in the nucleus is in fact equal to the number of electrons in the cloud that surrounds the nucleus, thus making the atom as a whole electrically neutral, since the charge on the proton is numerically equal to that on the electron. The protons and neutrons are collectively called *nucleons*, and are bound together in the nucleus by an extremely powerful force known as the *nuclear force*. Although the exact nature of this force still remains something of a mystery, its properties are now well established. Its range of influence is limited to the size of the nucleus, it is always attractive, and it acts in the same way between nucleons irrespective of whether they are charged protons or uncharged neutrons. It is the strongest known force in nature, and, as a result, the nucleons are confined to a very small volume, giving rise to a very high value for the density of the resulting nuclear matter. If nuclei could be packed closely together, a heaped teaspoonful would weigh as much as a million automobiles! Figure 1.2 shows a schematic representation of the structure of a typical nucleus.

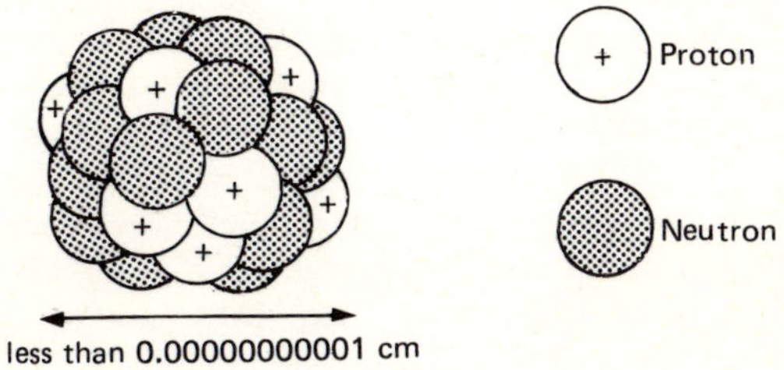

Figure 1.2 *The structure of a typical nucleus*

Isotopes

All atoms of a given element have the same chemical properties, and therefore the same number of electrons orbiting their nuclei. Since the atom is electrically neutral, all atoms of a given element must therefore also have the same number of protons in their nuclei. The number of neutrons need not be the same, however, and this results in different forms of an element which are known as *isotopes*. Uranium, for example, which has 92 protons in its nucleus, consists mainly of two isotopes, namely *uranium 238* (which has 146 neutrons in its nucleus) and *uranium 235* (which has only 143). Natural uranium is a mixture of roughly

99.3 per cent uranium 238 and only 0.7 per cent uranium 235 – the isotope that is used as a fuel for nuclear reactors.

All elements have isotopes. The element hydrogen, for example, has three – known as *hydrogen, deuterium* and *tritium*. Hydrogen has a nucleus which consists only of a single proton, but the deuterium nucleus contains a neutron as well. Since the neutron and proton masses are about the same, a deuterium atom weighs roughly twice as much as a hydrogen atom; possibly because of this, deuterium is often referred to as 'heavy hydrogen'. The third isotope, tritium (one proton and two neutrons in the nucleus), is radioactive, and we shall meet it again later, when the possibility of generating electricity using nuclear fusion is explored. Figure 1.3 shows the structures of the nuclei of the three isotopes of hydrogen.

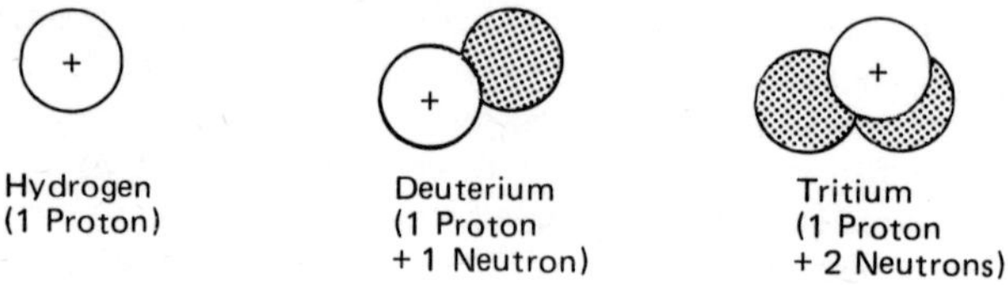

Figure 1.3 *The nuclei of the isotopes of hydrogen*

Nuclear binding energy

In their show *At the Drop of Another Hat*, the British entertainers Flanders and Swann told a story relating some of their experiences when flying across the Atlantic. At one point, Michael Flanders recounted the problems of trying to keep within the airline baggage allowance of 20 kilogrammes. He weighed his suitcase empty – 5 kilogrammes, and his clothes – 15 kilogrammes, filled his case and weighed it again – 21 kilogrammes! It appeared that the whole was greater than the sum of the parts. If the nucleus behaved in this way, there would be no nuclear power industry, and, incidentally, no atomic bombs.

With the nucleus, the sum of the parts is greater than the whole; in other words, the total mass of the individual protons and neutons is greater than the mass of the nucleus formed from them. When protons and neutrons come together to form a nucleus, part of the total mass is converted into energy – the energy that binds the nucleus together. This energy is known as the *binding energy*, and its existence is further demonstrated by the fact that exactly the same amount of energy must be applied to the nucleus to overcome the nuclear forces and pull it apart. The idea of the interconversion of mass and energy is known as *mass-energy equivalence*, and was first expressed mathematically by Einstein. The formula is actually a very simple one: $E = mc^2$, where E is energy, m the change in mass and c the speed of light (300 million metres per second). The change in mass which occurs when a nucleus is formed from protons and neutrons is known as the *mass defect*, and, in this case, E is the binding energy. The 'energy equivalence' of mass can best be illustrated in the following example:

> *If 1 kilogramme of mass were to be totally 'annihilated', it would result in the 'release' of about one hundred thousand million million joules* of energy (normally written as 10^{17} J). This enormous amount of energy is sufficient to boil the water dry in roughly 200,000 Olympic swimming pools!*

In nuclear physics, however, the mass defects are *very much smaller* than this, and, as a result, the energy required to pull a nucleus apart is also correspondingly smaller. In practice, the joule is far too large a unit for energy calculations in nuclear physics, and scientists therefore use a smaller unit known as the *electron volt* (or *eV*) instead. This is a very tiny unit indeed, over six million million million being needed to make one joule. Sometimes, a larger unit equal to one million electron volts (written 1 MeV in scientific shorthand) is used for convenience.

If we plot the *average binding energy per nucleon* (the total binding energy of the nucleus divided by the total number of protons and neutrons) against *mass number* (the total number of protons and neutrons), we obtain the graph shown in Figure 1.4 overleaf. This is an extremely important graph, and it is because of its shape that it is possible to extract energy from the nucleus.

How energy can be released from the nucleus

Looking carefully at the shape of the binding energy curve in Figure 1.4, it is noticeable that the curve rises to a maximum at a mass number of around 60. This corresponds to isotopes of elements such as iron, cobalt, nickel and copper. The high value of the total binding energy at this point indicates that the nuclei of these elements are more stable than those of either the very light elements (such as hydrogen) or the very heavy elements (such as uranium) at the extreme ends of the curve. Since all physical systems release energy when they move from a less stable to a more stable state, the curve indicates that there are *two* distinct ways by which energy can be liberated from the nucleus.

In the first, known as *nuclear fission*, the nuclei of a very heavy element are split up to form lighter nuclei, the difference between the initial and the final values of the total binding energy being released in the process. In the second, known as *nuclear fusion*, very light nuclei are joined together to form heavier nuclei – again with a net release of energy. In both cases, the arrows shown in Figure 1.4 indicate that the binding energy per nucleon at the end of the process is greater than at the beginning. Fission and fusion are both capable of releasing enormous amounts of energy – roughly one million times greater than that released by chemical

* The *joule* is the standard unit of energy used in science; it is the amount of energy needed to lift a mass of one kilogramme through roughly ten centimetres.

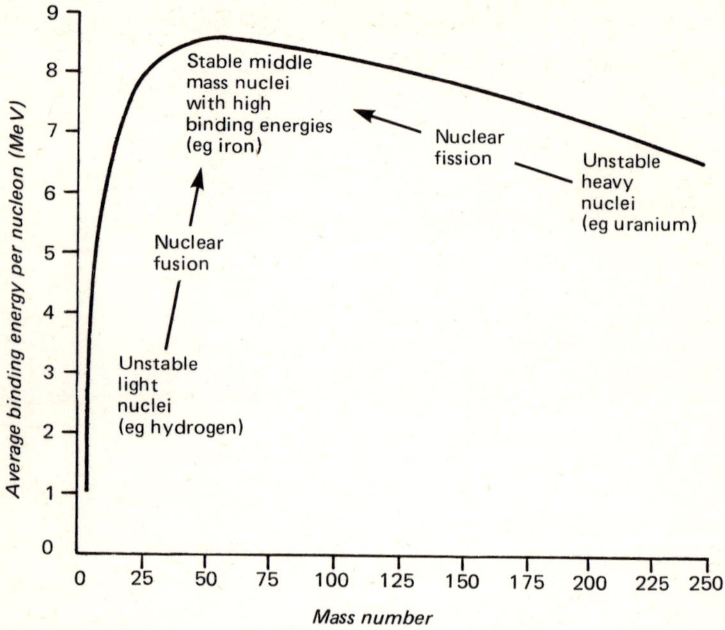

Figure 1.4 *Binding energy per nucleon plotted against mass number*

reactions between atoms and molecules. Each fission of a uranium nucleus, for instance, releases about 200 MeV of energy. If you consider that one kilogramme contains roughly 2.5 x 10^{24} (2.5 million million million million) nuclei, then, theoretically, the total fission of this amount of uranium would be accompanied by an energy release of about 10^{14} (100 million million) joules – enough to boil 200 of our Olympic swimming pools dry!

Nuclear fission and its applications

Nuclear fission is the process by which nuclear weapons of the type commonly known as *atom bombs* work, and is also the process by which all *nuclear reactors* operate. Figure 1.5 shows the details of the fission process in schematic form, taking U235 as an example. (As we shall see in Chapter 2, it is the nuclei of U235 that form the basis of all our commercial nuclear power stations at present.) The U235 nucleus is relatively unstable, and, when it is struck by a neutron, the energy of the neutron is sufficient to split the uranium nucleus into two roughly equal parts, releasing two or three further neutrons and a large amount of energy in the process. The two new nuclei, or *fission fragments* as they are called, have binding energies much closer to the maximum point of the curve shown in Figure 1.4 than the original uranium nucleus.

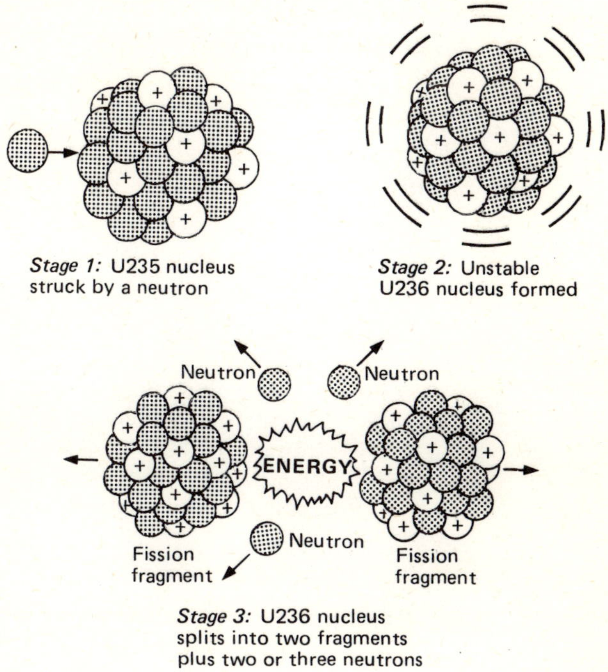

Figure 1.5 *The fission of a uranium 235 nucleus*

The fission process can be compared with the behaviour of a large drop of liquid resting on a tabletop. Surface tension forces hold the molecules of the liquid together in drop form, but can easily be upset if the surface of the drop is disturbed. Hitting the drop gently with a pencil or similar implement will often rupture the drop and split it into two or more smaller drops. In our analogy, the large single drop is the uranium nucleus and the pencil is the neutron. The laws of physics tell us that a small drop is more stable than a large drop of the same liquid. The binding energy per nucleon is analogous to the surface tension forces, and Figure 1.4 tells us that the middle-mass nuclei (the fission fragments in this case) are more stable than the heavy nucleus (the original uranium 235 nucleus). The liquid-drop analogy is demonstrated pictorially in Figure 1.6.

The principal difference between an atomic bomb and a nuclear reactor is that in the former, the rate at which fission takes place is uncontrolled, so that virtually all the fissile nuclei in the core of the bomb undergo fission in a small fraction of a second, releasing all their energy in the

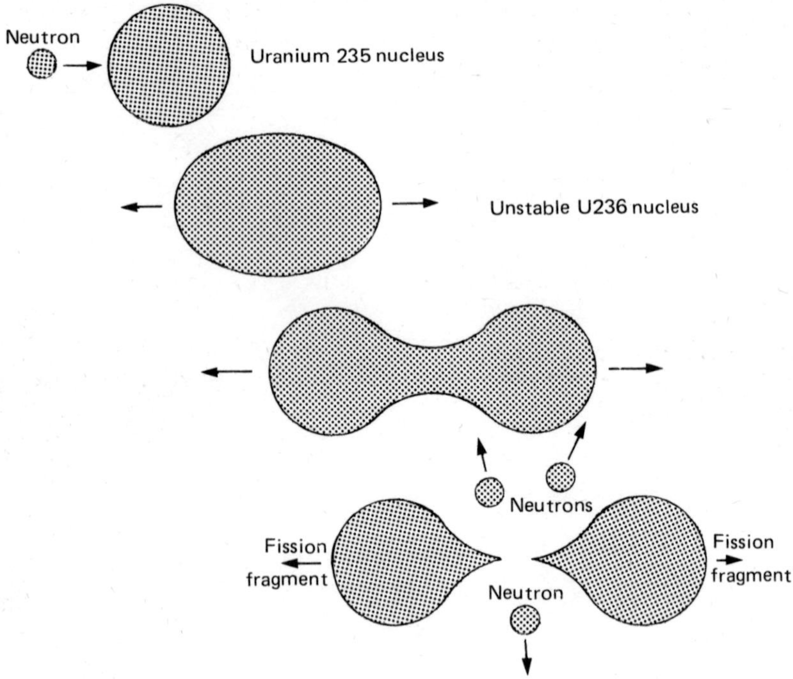

Figure 1.6 *The liquid-drop model of fission*

process (hence the explosion). In the latter, the fissile nuclei release their energy at a steady, controlled rate. How it is possible to use the fission process in such dramatically different ways will be explained in the next chapter.

Nuclear fusion and its applications

Nuclear fusion is the process by which the sun produces its energy and by which thermonuclear weapons (commonly known as *hydrogen bombs*) work. Both convert hydrogen into helium, releasing large amounts of energy in the process. In principle, it should be possible to build a power station that produces electricity by harnessing the energy released in thermonuclear reactions. One fusion reaction which is being investigated with this purpose in mind is shown in Figure 1.7. Here, nuclei of the isotopes deuterium and tritium 'fuse' together to form a Helium 5 nucleus, which is so unstable that it immediately changes to Helium 4, emitting a neutron and releasing its excess energy in the process.

A large amount of basic research is being carried out on this reaction, but the day when nuclear fusion will be able to meet all our electricity

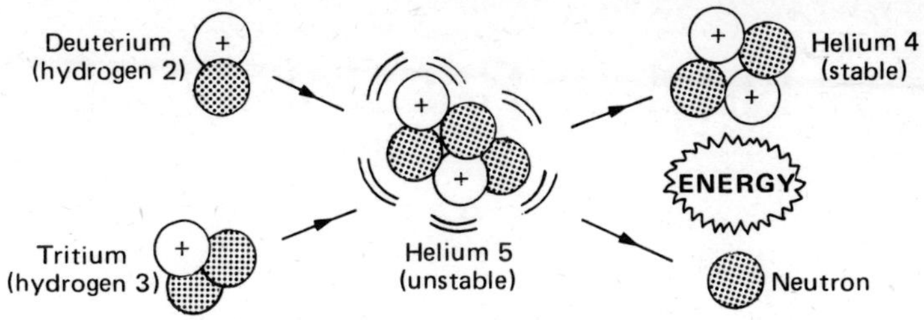

Figure 1.7 *The deuterium-tritium thermonuclear (fusion) reaction*

needs still seems to be a very long way off. There are many technical difficulties associated with the harnessing of nuclear fusion for peaceful purposes, and we will look at these in some detail in the latter part of Chapter 2.

CHAPTER TWO

How Nuclear Reactors Work

In Chapter 1, we saw how it is theoretically possible to persuade the atomic nucleus to give up part of its energy by nuclear fission and by nuclear fusion. In this chapter, we will look at the ways in which these sources of energy can be harnessed in nuclear power stations.

We will begin by examining fission reactors – the only type that has been successfully built so far. After pointing out the similarities between nuclear and conventional power stations, we will explain how it is possible to release fission energy from nuclei at a steady, controlled rate. Then, we will describe the basic features of fission reactors and examine the main types and designs currently in use. Finally, we will return to the subject of nuclear fusion in order to see how it is hoped to harness this potentially limitless source of energy.

Nuclear power stations as energy converters

The first thing to be appreciated about nuclear power stations is that they are in many ways identical to conventional power stations. Both are in effect gigantic *energy converters*, using the energy contained in their primary fuel to produce heat and raise steam which is used to drive turbines which, in turn, drive the generators that actually produce the electricity. As can be seen from Figures 2.1(a) and (b), the only real difference between nuclear and conventional power stations is the way in which the heat is produced to generate the steam.

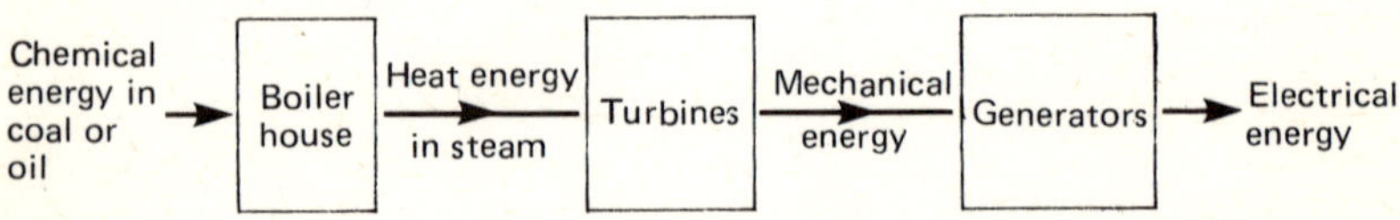

Figure 2.1(a) *The various energy conversion processes that take place in a conventional power station*

In a conventional power station, the heat is produced by releasing the chemical energy contained in fossil fuels such as coal or oil. This is done

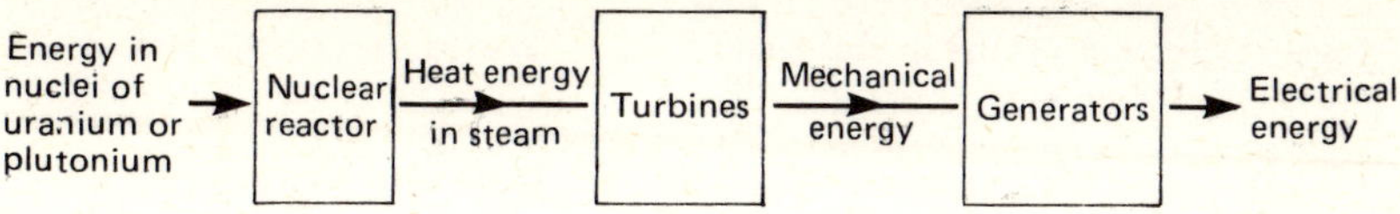

Figure 2.1(b) *The various energy conversion processes that take place in a nuclear power station*

by burning the fuel in a boiler house (in principle, simply a much larger version of the household central heating boiler). In a nuclear power station, the heat is produced by releasing the energy contained in uranium or plutonium nuclei by making them undergo controlled fission in a nuclear reactor.

Nuclear chain reactions

For energy to be released at a steady, controlled rate from uranium or plutonium, it is first necessary to arrange for the nuclei of the atoms to undergo fission in a continuous manner. In Chapter 1, we explained how a neutron can cause fission of a uranium or plutonium nucleus, and, in so doing, release (on average) two or three further neutrons. These fission neutrons can be used to cause further fissions of uranium or plutonium nuclei, and so perpetuate the process by what is known as a *chain reaction.* If more than one fission neutron in each generation go on to cause further fissions, the number of fissions will increase rapidly with time, and an explosion may ultimately take place. This is an example of an *uncontrolled chain reaction* of the type that occurs when an atomic bomb is exploded, and is illustrated schematically in Figure 2.2.

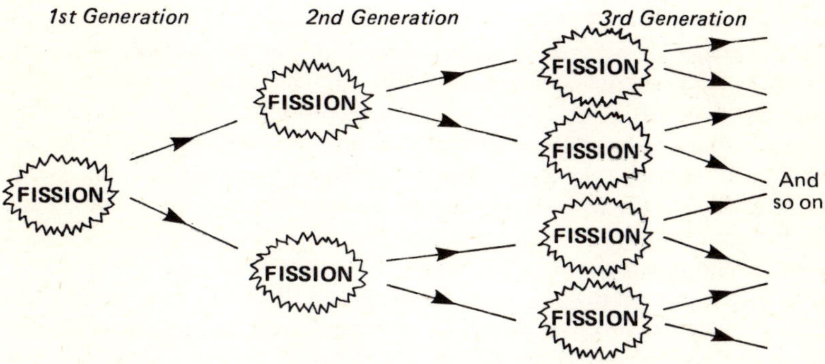

Figure 2.2 *An uncontrolled chain reaction of the type that takes place in an atomic bomb explosion*

If it is arranged that, in each generation, only *one* fission neutron is allowed to cause a further fission (as shown in Figure 2.3), the fissions take place at a constant rate. This means that the fissile fuel releases its energy in a controlled manner and no explosion can occur. In addition, this arrangement results in a *self-sustaining chain reaction* – the basis of all power-producing reactor designs.

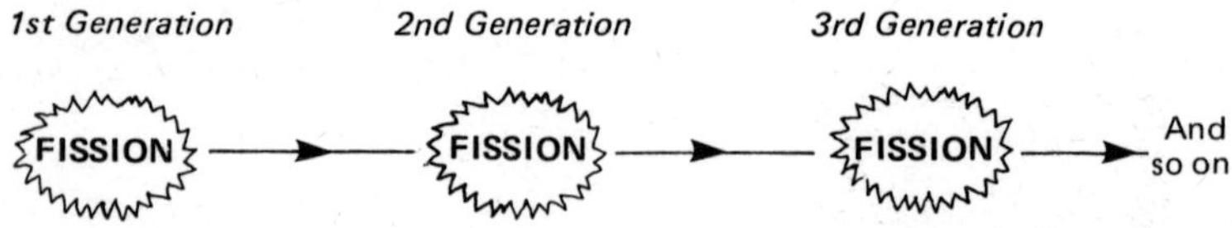

Figure 2.3 *A controlled chain reaction of the type that takes place in a nuclear reactor*

Nuclear reactors

There are two basic types of nuclear reactors, namely, *thermal reactors* and *fast reactors*. To date, only thermal reactors have been used in commercial nuclear power stations, although Britain, France, Germany, Japan, the USA and the USSR are all considering building their first commercial power-producing fast reactors. Since these two types of fission reactors operate on slightly different principles, we will first of all distinguish between them, and then move on to describe each in more detail.

The neutrons that are produced during the fission process are known as *fast neutrons*. They have a range of energies spread around a value of one MeV (see Chapter 1 for definition), and their speeds are of the order of 14 million metres per second (roughly 1/20 of the speed of light). As their name suggests, fast reactors use these fast neutrons to cause fission in the plutonium fuel that they use.

Thermal reactors, on the other hand, use uranium as fuel, and, in their case, it is only the U235 nuclei that undergo fission. Now U235 does not readily undergo neutron-induced fission unless the neutrons have a relatively low energy – about 1/40 eV in fact (this corresponds to a neutron speed of about 2200 metres per second). Thus, if fission neutrons are to be used to produce a controlled (and self-sustaining) chain reaction with U235 as fuel, they will obviously have to be slowed down. An energy value of 1/40 eV corresponds roughly to the energy of the air molecules we breathe, and is known as *thermal energy*. Fission neutrons that have been slowed down to the same energy are known as thermal neutrons, and it is for this reason that the reactors which use them are known as thermal reactors. The need to slow down the fission neutrons before allowing them to cause further fissions in the fuel of thermal reactors is the principal reason for the design differences between such reactors and fast reactors.

Thermal reactors

The essential features of a power-producing thermal reactor are shown in schematic form in Figure 2.4.

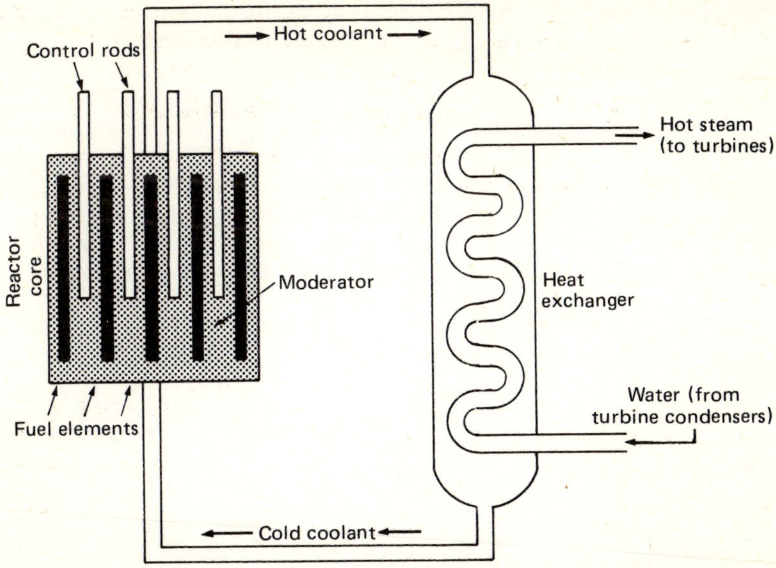

Figure 2.4 *Schematic diagram of a power-producing thermal reactor*

The heart of the reactor is the *core* – an array of fuel elements in which the fissile fuel is contained and in which the energy-producing chain reaction takes place. The fuel is confined in special corrosion-resistant canning (or *cladding*) materials so that it cannot move about in the core and contaminate other parts of the reactor such as the coolant. The fission neutrons are not, however, stopped by the canning material, and can therefore leave the fuel element in which they are created and react with fissile nuclei in other fuel elements. It is during their travels between fuel elements that the neutrons are slowed down to thermal energies by making them collide with the nuclei of a material known as a *moderator*. The geometrical arrangement of fuel and moderator in the core is designed so that the fast neutrons produced by fission are slowed down to thermal energies before they meet more fissile material in another fuel element.

As can be seen from Figure 2.4, the core also contains *control rods*. These are rods of neutron-absorbing material which are used to regulate the rate at which heat is generated within the core. For a reactor working at a given power level, the control rods absorb any excess neutrons produced so that a self-sustaining chain reaction is just maintained. To *increase* the power output of the reactor, the rods are gradually moved *out* of the core, and to *decrease* the power output they are gradually moved further *into* the core.

A further feature, not shown in the diagram, is the presence of *shut-*

down rods. These are similar to the control rods in that they strongly absorb neutrons, but are used only to start up or shut down the reactor. When the shut-down rods are inserted into the core, they absorb so many of the fission neutrons that a self-sustaining chain reaction is impossible. Thus, the power level rapidly falls to zero and the reactor is shut down. When the rods are taken out of the core, the neutron population can increase, so that the power level increases; it can then be set at the required value by adjusting the positions of the control rods. The shut-down rods therefore operate on the reactor rather like an on/off switch, while the control rods function rather like the dimmer control on a room light.

The energy that is released in a power-producing reactor is removed by means of a circulating *coolant* – generally a high-pressure gas, or water. This is heated up as it flows through the core, and is then passed through *heat exchangers*, where it transfers its heat energy to water which is boiled to produce steam. Because the coolant is under high pressure, the core has to be surrounded by a *pressure vessel* of some kind. This is usually made of steel or pre-stressed concrete. The pressure vessel is, in turn, totally enclosed within a *safety vessel*. The core becomes highly radioactive once the fuel begins to produce fission fragments, so power workers have to be protected from the resulting radiation by thick layers of reinforced concrete and stainless steel. The protective outer layers that surround a nuclear reactor are sometimes referred to as the *biological shield*.

These, then, are the basic structural features of a thermal reactor. Before we move on to the fast reactor, we will take a look at some specific types of thermal reactors in order to see how these features have been built into their design.

Some thermal reactor designs

1. Gas-cooled reactors: (a) The Magnox reactor (UK)

Magnox reactors were developed in Britain in the 1950s, where they were used in the first generation of nuclear power stations. In all, 13 Magnox stations were built, including one in Italy and one in Japan. The reactors were named after the material in which their fuel is contained – an alloy of magnesium.

Magnox reactors use natural uranium fuel in the form of metal rods encased in a magnox sheath, each 12-kilogramme rod producing as much electricity as 150 tonnes of coal when 'burned' in a reactor. The moderator is a form of carbon known as graphite, and the core simply consists of a large number of graphite blocks in which holes have been drilled in order to accommodate the fuel elements, control and shut-down rods, and instrumentation. The coolant used is high-pressure carbon dioxide gas, which is pumped through the holes in the core. A schematic diagram of a Magnox power station is shown in Figure 2.5, and photographs of a typical Magnox station are shown in Figures 2.6(b) and 8.1.

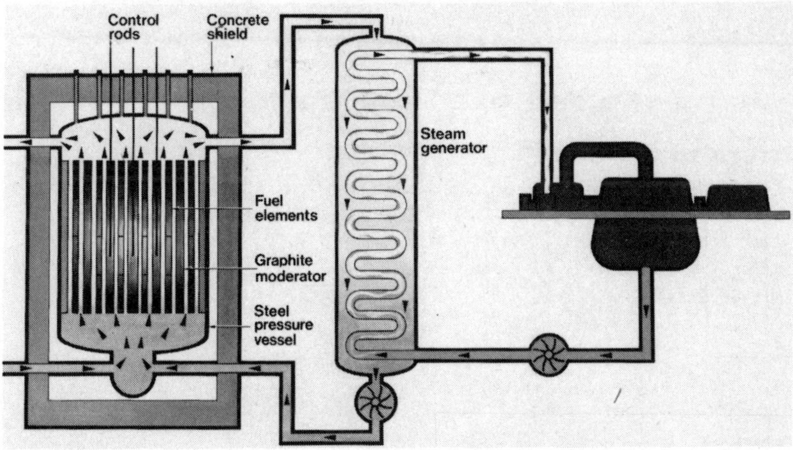

Figure 2.5 *A schematic diagram of a Magnox power station*

(b) The advanced gas-cooled reactor – AGR (UK)

The thermal efficiency of a power station, ie the electrical output per unit of thermal (heat) energy input, depends upon the temperature of the steam that is used; the higher the steam temperature, the greater the efficiency. The AGR was designed in order to improve the comparatively low efficiency that was possible using the Magnox system by increasing the steam temperature. To cope with these higher temperatures, the original metal fuel was replaced by a ceramic oxide fuel, and the magnox alloy casing was changed to stainless steel. Unfortunately stainless steel absorbs neutrons rather more easily than magnox alloy, and, to overcome this extra 'loss' of neutrons, the proportion of fissile nuclei (U235) had to be increased from its natural level of 0.7 per cent to about two per cent by a process known as *enrichment* (see Chapter 3). Because of the higher coolant pressure, the steam generators (heat exchangers) had to be located inside the pressure vessel rather than outside as in the case of Magnox stations. This is shown clearly in Figure 2.6(a), a schematic diagram of an AGR power station. Like the Magnox reactor, the AGR uses graphite as a moderator and carbon dioxide as a coolant.

A total of eight AGR stations have so far been built or are under construction in Britain. A photograph of one of these stations is shown in Figure 2.6(b).

(c) The high-temperature gas-cooled reactor – HTR

This reactor type, which is under consideration by several countries as a possible commercial power system, represents the possible next generation of gas-cooled reactors. It is designed to operate at even higher temperatures and pressures than the AGR. Graphite is still used as the moderator, but

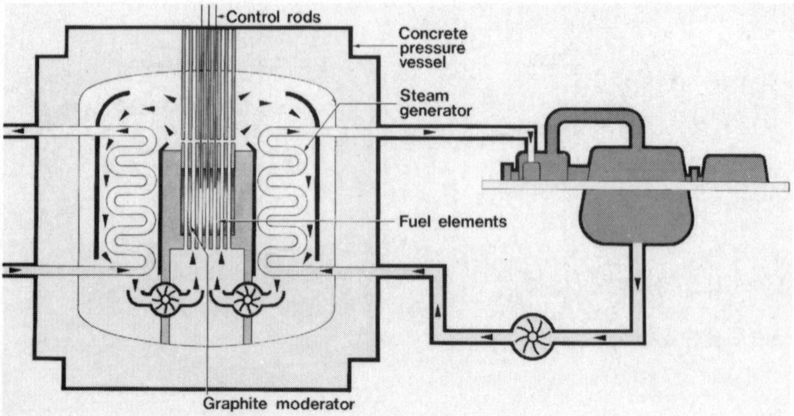

Figure 2.6(a) *A schematic diagram of an AGR power station*

Figure 2.6(b) *A typical AGR power station — the 1200-MWe Dungeness 'B' Power Station in Kent, England (Dungeness 'A', an earlier Magnox station, is seen in the background)*

the fuel is now in the form of highly enriched uranium oxide spheres (containing ten per cent of U235) encased in a carbide coating. These are bonded into a graphite matrix and assembled as fuel elements. In this form, the fuel can operate at higher temperatures than metal-clad fuel, and a larger proportion can be 'burned up' in each fuel cycle. The carbon dioxide gas used as a coolant in the Magnox and AGR designs would corrode the graphite moderator at the high temperatures prevailing in the HTR, so helium gas (which is completely inert) is used instead.

One of the attractions of the HTR is that it operates at such high temperatures that it might be possible to replace the steam generator and turbine by a gas turbine, or even to use the heat directly in industrial processes.

2. Light water reactors

It is probably fair to say that most American power-producing reactor systems are now based on the light water* reactor because of the early needs of the American nuclear submarine programme, which had to use water-cooled reactors for technical reasons. The total power-producing capacity of the various light water reactors that have so far been built all around the world is at least ten times that of gas-cooled reactors, which are almost entirely confined to the UK.

Under normal atmospheric pressure, water boils at 100°C. Steam at this temperature (and pressure) would never drive a large turbine, so, in light water reactors (LWRs), the boiling point of the water is raised by pressurizing the system. LWRs are of two types, the *pressurized water reactor* (PWR) and the *boiling water reactor* (BWR). In the PWR, the water does not actually boil, but gives up its heat in a heat exchanger. In the BWR, the pressure is lower (although still high) and the water is allowed to boil, the resulting steam being used directly to drive the turbines.

(a) The pressurized water reactor – PWR (USA)

The PWR uses ordinary (light) water, as both a coolant and a moderator. Because ordinary water absorbs neutrons to a significant extent, the uranium oxide fuel has to be enriched until it contains between three and four per cent U235. The ceramic oxide fuel is clad in an alloy of zirconium called 'zircaloy', and the core is contained in a steel pressure vessel with walls typically 15 to 20 centimetres thick. The coolant/moderator is heated by the core and pumped at high pressure to steam generators, where it boils water in separate circuits to produce steam. Like the AGR, the core, coolant circuits and steam generators are all contained within a thick concrete safety shield.

As stated above, most of the nuclear power stations so far built in the USA use the PWR system, and the third generation of British nuclear

* 'Light' (naturally occurring) water contains virtually no 'heavy' hydrogen (deuterium) in its molecules; in so-called 'heavy' water, on the other hand, the hydrogen is almost entirely replaced by deuterium (see Figure 1.3 in Chapter 1).

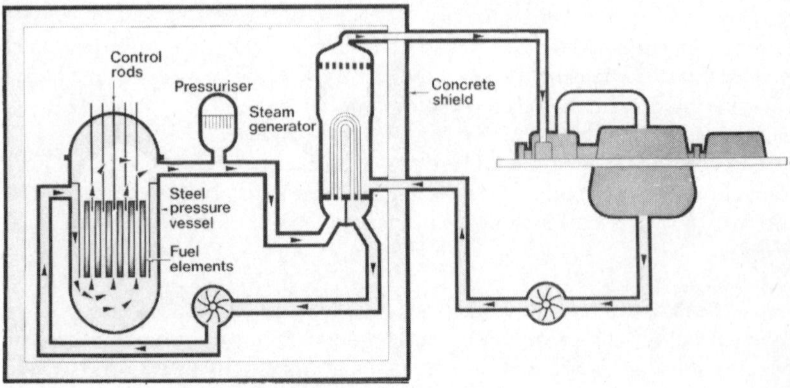

Figure 2.7(a) *A schematic diagram of a PWR power station*

Figure 2.7(b) *A typical PWR power station – the 575-MWe Connecticut Yankee Nuclear Power Station at Haddam Neck, Connecticut, USA*

power stations, which are to be built during the 1980s and 1990s, will probably include some using the PWR system. Figure 2.7(a) shows a schematic diagram of the PWR system, while Figure 2.7(b) shows a typical PWR power station.

(b) The boiling water reactor – BWR (USA)
As mentioned above, the principal difference between a BWR and a PWR is the pressure of the coolant/moderator system. In the BWR, no heat exchanger is needed, since the steam is produced directly from the coolant/moderator water. The basic design data for a typical 600-megawatt electrical output (MWe)* reactor of each type are compared in Figure 2.9.

3. Heavy water reactors
Heavy water, in which the hydrogen atoms of ordinary (light) water have been replaced by deuterium atoms, is expensive to produce, but has considerable advantages over ordinary light water when used as a moderator in a thermal reactor. Unlike the latter, heavy water is a very poor absorber of neutrons, thus allowing a heavy water-moderated thermal reactor to use natural uranium rather than enriched uranium as its fuel. There are two basic heavy water reactor designs – the Canadian *deuterium, natural uranium* (or CANDU) reactor and the British *steam-generating heavy water reactor* (or SGHWR).

(a) The CANDU reactor (Canada)
The general layout of a CANDU power-producing reactor system is shown in Figure 2.8(a). The fuel is natural uranium dioxide clad in zircaloy cans, with heavy water being used both as the moderator and as the coolant. The moderator is contained in a special tubed tank called a *calandria*, which is kept at normal atmospheric pressure and at a comparatively low temperature. The thin pressure tubes in the calandria contain the fuel elements. The heavy water coolant is pumped at high pressure through these tubes, where it is heated by the energy released from the fuel; the coolant is then pumped to steam generators, where it boils ordinary light water in separate circuits.

A photograph of a typical CANDU power station is shown in Figure 2.8(b).

(b) The steam-generating heavy water reactor – SGHWR (UK)
In this British design of reactor, the coolant is ordinary light water, which is allowed to boil and produce steam directly. As in the CANDU system just described, the coolant passes through a set of pressure tubes in the calandria which contains the fuel elements. The fuel in the SGHWR is enriched to about 2.25 per cent U235 (necessary because ordinary water is

* The *megawatt*, equal to one million watts, is the unit generally used to measure the power output of nuclear reactors. A reactor with a 600-MWe (megawatt electrical) output would be able to run roughly 600,000 single-bar electric fires.

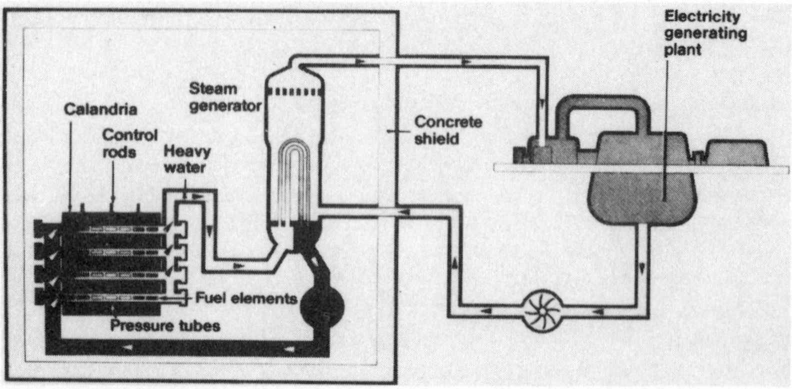

Figure 2.8(a) *A schematic diagram of a CANDU power station*

Figure 2.8(b) *A typical CANDU power station — the 600-MWe Gentilly-2 Power Station being built in Quebec, Canada (Gentilly-1, a prototype CANDU station, is seen in the background)*

used in the core) but otherwise takes much the same form as in the Canadian reactor. The heavy water moderator is also held in the calandria. The use of thin pressure tubes avoids the need for a large pressure vessel of the type used in the PWR design, thus making CANDU and SGHWR reactors easier to refuel and service.

Although a prototype SGHWR has been operating reliably at Winfrith in the UK for some years, the British government has now decided that its future thermal nuclear power programme will be based on the AGR and PWR designs, and earlier plans to introduce the SGHWR into commercial service have been cancelled (see Chapter 8).

In order to help you compare the different reactors described above, a table summarizing their main features is shown in Figure 2.9.

Fast reactors

Despite what many people think, fast reactors are not new. The first nuclear reactor of any kind to produce electricity in the USA was a fast reactor, and the original fast reactor built in Scotland at Dounreay in Caithness was producing electricity before any of the commercial Magnox stations ordered by Britain's electricity generating boards (see Chapter 8).

Since fast reactors use fast neutrons to produce fission of their fuel, they do not require a moderator. They also differ from thermal reactors in a number of other important respects.

Firstly, they use a much richer fuel, consisting of either plutonium or a mixture of plutonium and uranium in the form of ceramic oxide pellets. Plutonium is an artificial element (ie it does not occur naturally) that is produced in all nuclear reactors as a side product, being formed when neutrons are absorbed by uranium 238 nuclei.

Secondly, the fact that their fuel is very much richer than that used in thermal reactors means that, for a given power output, less of it is required. Thus, the core of a fast reactor is considerably smaller than that of a thermal reactor of comparable power. This means that it is no longer possible to remove the enormous amounts of heat that are generated in the small volume of the core using water or gas as a coolant. Instead, a liquid metal coolant must be used, since this is capable of removing heat from the core much more efficiently. The coolant generally chosen for fast reactors is liquid sodium, which flows freely at high temperatures and can pass easily through the small spaces between the fuel elements.

Thirdly, the use of liquid sodium as a coolant makes it necessary to employ two separate coolant circuits, since the sodium that flows through the core becomes highly radioactive owing to neutron irradiation and thus cannot be allowed to pass out of the reactor vessel. A schematic diagram of a liquid metal-cooled fast reactor, showing the principal design features, is given in Figure 2.10.

Another important feature of fast reactors is that they can be designed in such a way that they actually produce *more* plutonium than they consume. This is done by surrounding the core with a blanket of U238,

Reactor	MAGNOX	AGR	HTR	PWR	BWR	CANDU	SGHWR
Country of origin	Britain	Britain	several	USA	USA	Canada	Britain
Fuel	natural uranium metal	uranium oxide (2.3% enriched)	uranium oxide or carbide (10% enriched)	uranium oxide (3.2% enriched)	uranium oxide (2.6% enriched)	uranium oxide (natural)	uranium oxide (2.25% enriched)
Cladding	magnox	stainless steel	silicon carbide	zircaloy	zircaloy	zircaloy	zircaloy
Moderator	graphite	graphite	graphite	ordinary water	ordinary water	heavy water	heavy water
Coolant	CO_2	CO_2	helium	ordinary water	ordinary water	heavy water	ordinary water
Coolant outlet temperature	400°C	650°C	720°C	317°C	286°C	305°C	272°C
Coolant pressure	285 psi (pounds per square inch)	585 psi	700 psi	2220 psi	1035 psi	1270 psi	885 psi
Approximate core dimensions	14 metres dia x 8 metres high	9.1 metres dia x 8.5 metres high	8 metres dia x 5 metres high	3 metres dia x 3.7 metres high	3.7 metres dia x 3.7 metres high	7.1 metres dia x 5.9 metres high	6.5 metres dia x 3.7 metres high
Fuel burn-up (in megawatt days per tonne)	4000	13,000	100,000	20,000	20,000	9000	21,000
Power density (in kilowatts per litre of core)	1	4.5	6	50-100	50-100	16	11

Figure 2.9 Characteristics of the major thermal reactor types (figures for a typical 600-MWe system)

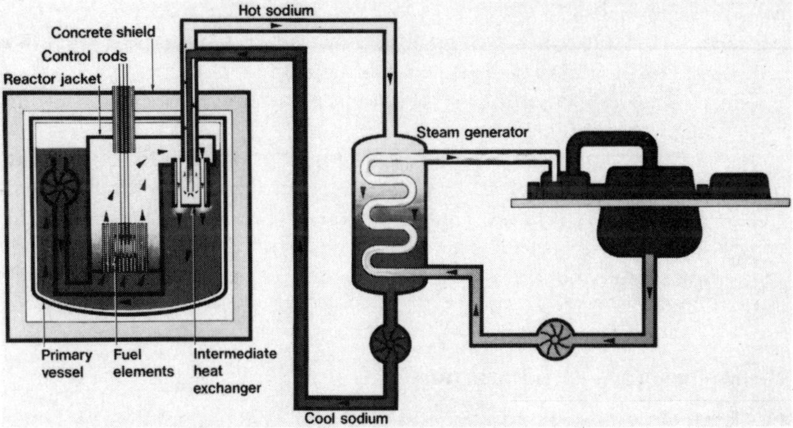

Figure 2.10 *A schematic diagram of a liquid sodium-cooled fast reactor power station*

which is gradually converted into plutonium by bombardment with neutrons escaping from the core. In this mode of operation, the reactor is sometimes referred to as a *fast breeder reactor*.

World interest in fast reactors

The world's first power-producing reactor was the one built at Dounreay in the 1950s. This was known as the *Dounreay fast reactor*, or DFR. DFR was followed by a second, larger reactor of 250 megawatts electrical output (intended as a prototype for a full-scale commercial reactor of roughly 1200 MWe) known as the *prototype fast reactor*, or PFR.

Work on fast reactors in continental Western Europe is now proceeding under a series of collaborative agreements between France, Germany, Italy, Holland and Belgium. Developments are probably most advanced in France, where work on fast reactors started in the 1960s with the construction of the test reactor *Rapsodie* at Cadarache. In 1973, a prototype power reactor, *Phénix* (250 MWe) was commissioned at Marcoule, and work on a 1200-MWe *Super Phénix* was begun at Creys-Malville in 1977. Super Phénix is being built as a joint venture with Italy and Germany, and is due to be completed in 1983.

Major fast reactor programmes are also under way in the USA, USSR, Japan and India. American interest dates back to the 1940s, and, over the next 20 years, the USA designed and built several facilities to test the feasibility of fast reactor systems. In 1980, however, the Carter administration postponed the 'commercialization' of fast reactors indefinitely, and the future of the US fast reactor programme is now being re-examined.

The first commercial-scale fast reactor to be built in the USSR, a 600-MWe station at Beloyarsk, is currently nearing completion. There are also plans to build a 1600-MWe fast reactor at the same site; this is due to be completed in 1989.

Japan already has one experimental fast reactor in operation, and plans to construct a 300-MWe prototype power station by 1986.

India's declared intention is to have fast reactors in operation by the beginning of the next century.

Perhaps the most important reason for all this interest in the fast reactor is that *up to 60 per cent* of the energy available in natural uranium can be extracted if it is used in a fast reactor fuel cycle. This should be compared with only about one to two per cent if the uranium is used solely in a thermal nuclear power programme. We shall return to this point in Chapter 9.

Fusion reactors – the prospects

In Chapter 1, we saw that it is theoretically possible to release energy from the nucleus by joining together (or *fusing*) very light nuclei to form heavier nuclei, and explained why this was so using Figure 1.4. This fusion process is, of course, the opposite of nuclear fission. With regard to the possibility of using the fusion process as the basis of a power-producing reactor system, it has two very important advantages over fission. Firstly, the raw materials from which the fuel for fusion reactors is derived (water and lithium) are available in virtually unlimited quantities, so that a fusion power programme would never be restricted by shortage or high cost of fuel. Secondly, fusion produces no highly radioactive waste products, so that a fusion power programme would have no associated waste management or waste disposal problems.* Here, for the moment, however, the good news ends. The bad news is that fusion is proving much more difficult to harness than it was at one time hoped, and, despite the massive amount of research that has taken place in (for example) Russia, America, Europe, Japan and China, it has not yet proved possible to develop a practical fusion reactor system. Thus, a commercial power programme based on the fusion process still appears to be a very long way off.

The conditions required for controlled thermonuclear fusion

In order to make the nuclei of light atoms fuse together, it is necessary to increase their thermal energy (ie to heat them up) so that they collide at very high relative velocities. At the moment, the most promising reaction is the one shown in Figure 1.7, where the two 'heavy' isotopes of hydrogen, deuterium and tritium, combine to form helium. This reaction occurs spontaneously if a mixture of the two materials is heated to a temperature of the order of 100 million degrees. At such temperatures, matter exists in what is known as the *plasma state*, a plasma being a gas in which the atoms have been stripped of some (or all) of their electrons.

For the practicable release of fusion energy, the rate at which fusion

* Apart from the dismantling and disposal of the actual reactor at the end of its useful life.

takes place must be such that the energy produced by the fusion reaction is greater than the energy input to the plasma. In order to achieve this situation, it has been calculated that we need to produce a plasma in which the product of the *particle density* (the number of particles per unit volume) and the *energy containment time* (the time for which the plasma is maintained at a temperature high enough for fusion to take place) is of the order of 2×10^{20} particle seconds per cubic metre or more. Mathematically, this condition, which is known as *Lawson's criterion*, is written as

$$n\, t_E > 2 \times 10^{20} \text{ particle seconds per m}^3$$

where n is the plasma particle density and t_E is the energy containment time.

Since n and t_E are multiplied together, it is obvious that if one is made very large, the other need not be quite so large for the critical condition to be satisfied. Two possible working regimes are currently being investigated. The first, known as *magnetic confinement*, involves producing a plasma of moderate particle density ($\sim 10^{20}$ per cubic metre) and confining it for roughly one second using powerful magnetic fields. The second, known as *inertial confinement*, involves producing a plasma of much higher particle density ($\sim 10^{32}$ per cubic metre) in which energy-producing fusion reactions will take place before the plasma has time to disperse, which it does in roughly 10^{-12} (ie 0.000000000001) seconds. Let us now examine these in greater detail.

Magnetic confinement

The principle underlying magnetic confinement is basically very simple. Essentially, it involves heating up a low-pressure gas to a high temperature by passing an extremely large electric current through it, and preventing the resulting plasma from dispersing by means of a suitably shaped magnetic field which deflects the charged particles back into the plasma when they attempt to diffuse outwards. High-temperature plasmas of the type used in fusion research cannot be contained using material walls, so such magnetic fields represent the only practical method of achieving artificial confinement.

During the last 25 years, a large amount of research on magnetic confinement has been carried out in various parts of the world. A variety of plasma and magnetic field geometries have been tried in an attempt to achieve the critical conditions described above and to overcome the intrinsic tendency of the resulting plasma to disperse or become unstable. These have included linear 'magnetic bottles' (in which the plasma is confined by making the field strongest at the two ends), cusp containment systems (of the type that results when similar magnetic poles of two powerful electromagnets are brought into proximity) and toroidal systems (in which the magnetic field and plasma have a 'doughnut' shape).

At the time of writing, the most promising configuration appears to be the so-called *tokamak* system, a type of toroidal system first developed in Russia at the Kurchatov Institute of Atomic Energy. This is shown schematically in Figure 2.11(a). It consists of a central toroidal chamber containing a low-pressure mixture of deuterium and tritium surrounded by a series of field coils wound round the main torus in such a way as to produce a toroidal magnetic field within its volume, ie a field whose lines of force are parallel to the central axis of the torus (see figure). The toroidal chamber is also traversed by a transformer core on one limb of which a primary coil is wound. When a large pulse of current is passed through the latter, this causes an extremely high current to be induced in the reaction chamber, parallel to the toroidal magnetic field. This heats up the plasma, and also produces a poloidal magnetic field, ie a field whose lines of force are circles at right angles to the axis of the torus (again, see figure). This poloidal field combines with the toroidal field to produce a helical (ie spiral) field that effectively prevents the plasma from escaping from the centre of the toroidal chamber.

Major tokamak research programmes are currently taking place in a number of countries, including the USSR, the USA, Japan, China and Europe. In Europe, it was realized that the cost and complexity of a project capable of assessing the potential of the tokamak as a means of producing reactor-grade plasmas was beyond the resources of any single country, so a decision was made to collaborate in a single joint project – the *Joint European Torus* or *JET Project*. A large tokamak machine is currently being constructed by a team of scientists and engineers from the EEC countries, Sweden and Switzerland on a site adjacent to the UKAEA's

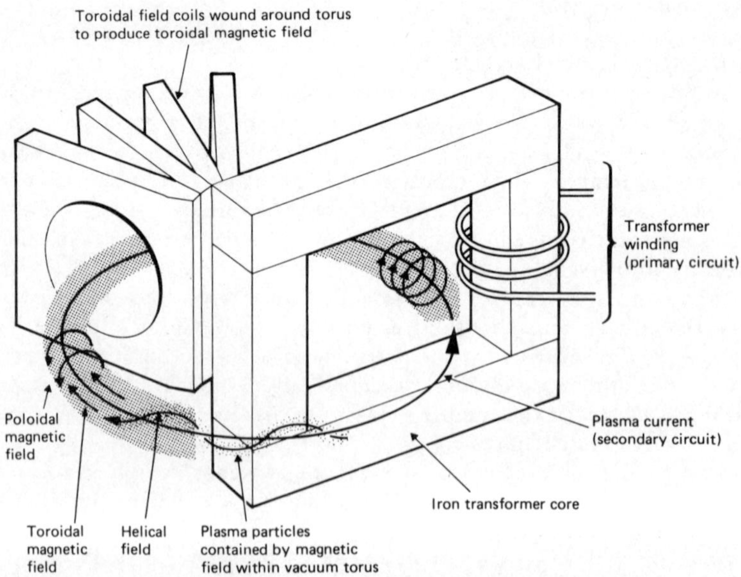

Figure 2.11(a) *A schematic diagram of a tokamak fusion reactor*

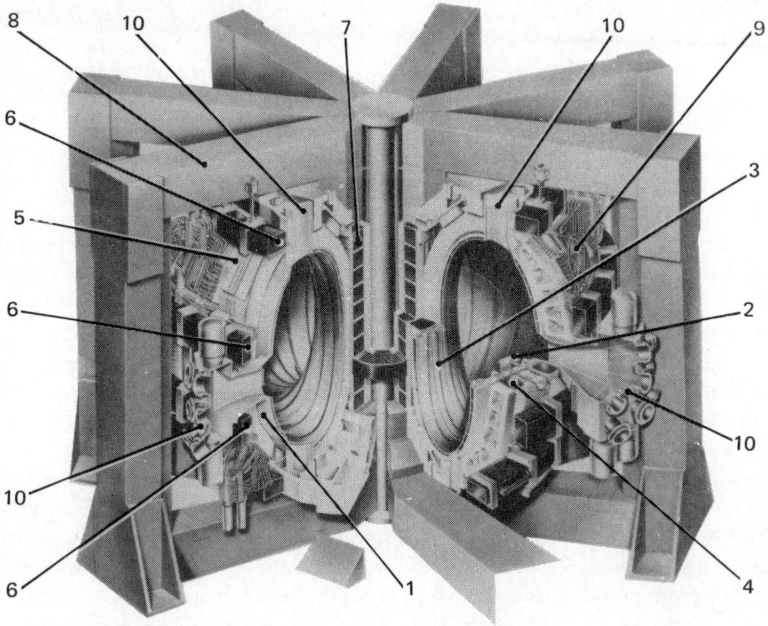

The JET apparatus
1. vacuum vessel (double walled)
2. material limiter defining the outer plasma edge
3. poloidal protective shields to prevent the plasma touching the vessel
4. toroidal field magnet of 32 D-shaped coils
5. mechanical structure
6. outer poloidal field coils
7. inner poloidal field coils (primary or magnetizing windings)
8. iron magnetic circuit (core and eight return sections)
9. water and electrical connections for the toroidal field coils
10. vertical and radial ports in the vacuum vessel

Figure 2.11(b) *The JET tokamak currently being built at Culham, England*

Culham Laboratory in Oxfordshire, England (see Figure 2.11(b)). It is hoped that this will achieve the critical thermonuclear conditions described above and provide useful information on the behaviour of deuterium-tritium plasmas. If the construction phase is completed on schedule, the experimental programme should begin in 1983.

Inertial confinement
The principle underlying the second containment system, inertial confinement, is even simpler than that underlying magnetic confinement. Here,

the object is to produce an extremely high-temperature plasma so rapidly that power-producing fusion reactions have time to take place before the plasma is dispersed by rising internal pressure. No artificial means are employed to confine the plasma, its natural inertia being used to hold it together for long enough for fusion to occur – hence the name of the system.

One method by which it is hoped to produce such a plasma is to concentrate an intense pulse of light from a concentrically focused bank of exceptionally high-powered lasers onto a pellet of lithium deuteride (or some similar material). By this means, it should (in principle) be possible to compress the fuel pellet to an extremely high density and to heat it to a sufficiently high temperature for a net release of energy to be produced. Lasers capable of generating the necessary power (about 10^{14} watts for 10^{-9} seconds) have not yet been developed, but appear to be feasible.

Future prospects

Thermonuclear fusion has now been demonstrated on a small scale in a number of laboratories, using both magnetic confinement and inertial confinement. In all cases, however, much more power was used to create the magnetic field, heat the plasma or operate the lasers than was liberated by the fusion reactions thus produced. The next stage will be to try to achieve break-even conditions, and, eventually, to produce useful amounts of power. Calculations suggest that this will require an extremely large

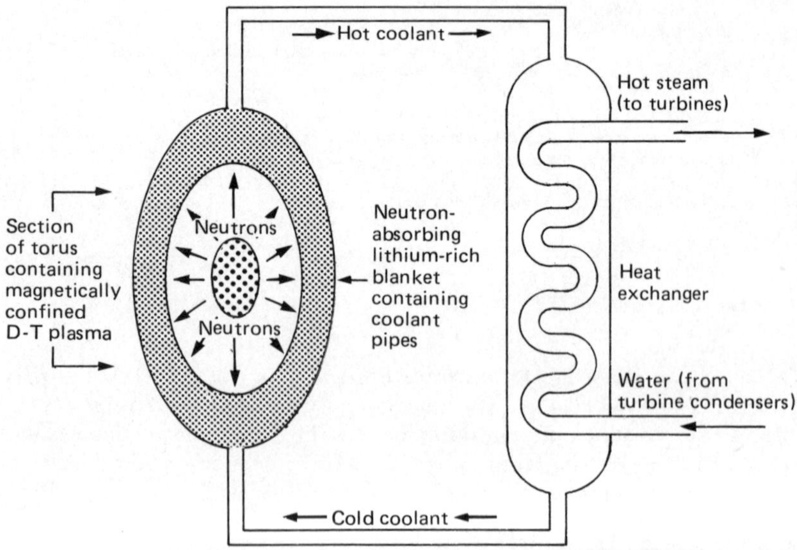

Figure 2.12 *Schematic diagram of a power-producing fusion reactor*

machine, and it is hoped that the large tokamaks currently being built in Russia, America, Europe and Japan will provide the data needed to design and build such a prototype fusion reaction. Figure 2.12 shows how such a machine might operate.

Since almost 80 per cent of the energy produced by the thermonuclear fusion of deuterium and tritium (the mixture envisaged as the fuel for the first generation of fusion reactors) will be carried off by the neutrons liberated during the fusion process (see Figure 1.7), it is planned to use these neutrons to carry energy out of the plasma. Being uncharged, the neutrons would pass freely through the magnetic field that contained the plasma before being stopped by a blanket containing lithium – a strong absorber of neutrons. The heat thus liberated in the blanket would then be carried off by a circulating coolant and used to raise steam – exactly as in the case of a fission reactor. Since lithium is gradually converted into tritium as it absorbs neutrons, such a reactor would also breed one component of its own fuel. The other component – deuterium – is available in unlimited quantities in sea water.

Even if the fusion experiments being planned for the middle and late 1980s prove to be completely successful, enormous engineering difficulties will still have to be overcome before a prototype power station can be built. Thus, it is not expected that fusion will become a practical source of energy in the foreseeable future – certainly not in this century. If all goes well, however, fusion could become a major source of energy sometime during the 21st century, and could eventually solve virtually all our energy problems.

CHAPTER THREE

The Nuclear Fuel Cycle

This chapter is devoted to a description of the life history of uranium when it is used as the basis of nuclear reactor fuel. We shall follow its progress from the mines where it is produced, to the fabrication plants where it is made up into fuel elements, to the reactors where it undergoes fission, and, finally, to the reprocessing plants where irradiated fuel elements are broken down and their contents either recycled or stored for future disposal.

The story is, however, complicated by a number of factors. In the USA, for example, the Carter administration called a halt to the reprocessing of all spent fuel elements in 1977 because of the fact that the use of nuclear power was growing worldwide, with many countries making plans for the use of plutonium in their civil nuclear power programmes. As a result of the President's concern over the possibility of further proliferation of nuclear weapons, the International Nuclear Fuel Cycle Evaluation Exercise (INFCE) was launched. Thus, the nuclear fuel cycle currently in operation in the USA is not, strictly speaking, a 'cycle' at all. As we shall see, there are in fact a number of alternative nuclear fuel cycles, all of which have their own special characteristics, and all of which have particular environmental and political implications. In this chapter, we will confine ourselves to describing the basic features of these various fuel cycles, thus preparing the way for detailed discussion of their wider implications in Parts III and IV. Before we do so, however we will try to answer the basic question: 'why do we have to reprocess nuclear fuel at all?'

Why reprocess nuclear fuel?

The reprocessing of nuclear fuel elements is perhaps the most important feature that distinguishes nuclear power stations from conventional power stations fuelled by oil or coal. There are two main reasons why reprocessing makes sense. Firstly, it is never possible to burn up all the fissile material in a given fuel element in a single irradiation cycle, largely because of the progressive build-up of fission products inside the fuel elements. Since the fission products occupy a greater volume than the original fuel, their creation causes the fuel element to swell. Thus, if the burn-up were to be taken beyond a certain limit, the fuel element would buckle and (eventually) rupture. This would constrict the coolant flow in the reactor and also contaminate the coolant itself. Since burn-up factors are normally very low, it is economically desirable to recover the unburnt fuel so that it can be recycled.

Secondly, *all* nuclear power stations produce plutonium as a side product. This material is used in the manufacture of fast reactor fuel and must therefore be separated from the spent fuel before it is available for this purpose. Plutonium can also be burned in thermal reactors, although, as we shall see later, this is not a very efficient way of making use of this valuable material.

Let us now look at the fuel cycles themselves, starting with the so-called *once-through cycle*.

The once-through fuel cycle

This is the simplest of all fuel cycles and is shown in schematic form in Figure 3.1.

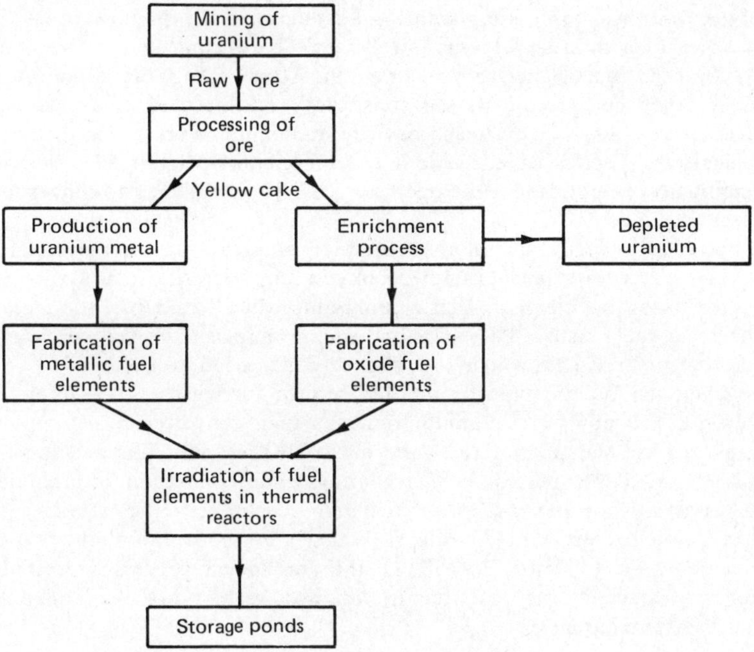

Figure 3.1 *The once-through fuel cycle*

The once-through fuel cycle is currently being used in the USA because of the moratorium on reprocessing that was imposed by the Carter administration, and is also used in Canada. From Figure 3.1, we see that, in this particular cycle, the plutonium is not recovered from the spent fuel elements. Instead, the complete fuel elements are stored indefinitely in large ponds.

The various stages of the once-through fuel cycle form the basis of all other fuel cycles, and so, at any future time, this cycle can be converted into any other cycle simply by breaking in at the storage pond stage and introducing a reprocessing facility. Let us now examine the successive stages of the cycle in detail.

Uranium production

The first stage in the cycle is the mining of uranium ore, which is found in large quantities in the USA, Australia, South Africa, Canada, Niger and Namibia. The mining methods used to extract the ore depend on the characteristics of the ore deposit. In general, open-cast methods are used if the deposits lie less than about 150 metres below the surface, while conventional underground mining techniques are used when the ore deposits lie at greater depths.

The uranium ores that are currently being mined generally contain between one and three kilogrammes of uranium to the tonne – usually in the form of a mixture of uranium oxides. This is equivalent to between 0.1 per cent and 0.3 per cent of uranium. After mining, the crude ore is first crushed and ground to the consistency of fine sand, and chemical solvents are then used to 'leach out' the uranium. The resulting uranium 'concentrate', often called *yellow cake*, contains roughly 80 per cent uranium by weight, and is shipped to a fabrication plant for conversion into nuclear fuel.

To be suitable for use as a reactor fuel, uranium must be particularly pure. It must not, for example, contain any significant amounts of neutron-absorbing elements that would reduce the efficiency of the fission process in the reactor. To remove all such impurities, the yellow cake is therefore first passed through a number of purification processes.

Depending on the type of thermal reactor for which the uranium is intended, the now-pure uranium oxide is then converted into uranium metal (for the Magnox reactors used in the UK), uranium dioxide (for the Canadian CANDU reactors) or into a gaseous compound of uranium known as *uranium hexafluoride* (for enrichment prior to the manufacture of the enriched uranium dioxide fuels that are used in most other types of reactor – see Figure 2.9). Since the enrichment process is a vitally important stage in the manufacture of most reactor fuel, we will now examine its main features.

The enrichment process

The fuels required for most thermal reactors, eg the advanced gas-cooled reactor (AGR), the pressurized water reactor (PWR) and the boiling water reactor (BWR), have to have their U235 content artificially increased above its natural level of 0.7 per cent for technical reasons (see Chapter 2). The varying degrees of enrichment required can be achieved in a number of ways, but the main commercial processes currently in use are *gaseous diffusion* and the *gas centrifuge*.

(a) Gaseous diffusion

The gaseous diffusion process was originally developed for the large-scale production of highly enriched uranium for use in the manufacture of atomic weapons. The plants built for this purpose have also been the main source of enriched uranium fuel for nuclear reactors, however, despite the fact that such reactors require their fuel to be enriched to only two to three per cent U235. Gaseous diffusion plants are very large (covering roughly half a square kilometre) and require enormous amounts of power to run them (typically 2000 megawatts of electricity – enough to supply two cities each the size of Glasgow, Scotland).

When a gas passes through a porous material (or *membrane*) its flow rate depends upon the mass of the molecules of the gas – the heavier the molecules, the slower they pass through the membrane. If the gas contains molecules with different masses, passing the gas through such a membrane will therefore bring about partial separation of the different components, producing a gas with a greater proportion of the lighter molecules. The gas used in the enrichment of uranium is the uranium hexafluoride (or 'hex', as it is called) that is mentioned above. On passing through a membrane, the outgoing gas is slightly richer in U235 than the ingoing gas, one pass increasing the U235 content by about 0.4 per cent. To achieve the two to three per cent U235 levels required for nuclear fuels, the process has to be repeated many times in a series of *cascades*, where the slightly enriched gas from one stage is fed into the next stage and so on. The need for many such stages is the principal reason for the large plant size.

Uranium hexafluoride is a particularly suitable choice of plant feed. It is a gas at comparatively low temperatures, and the fluorine gas that is used in its preparation consists of atoms of only a single isotope. The difference in mass of the uranium hexafluoride molecules is therefore due only to the different masses of the uranium atoms (U238 and U235).

The 'waste product' from a gaseous diffusion plant is uranium hexafluoride that has *less* than the natural level of U235 (typically only 0.3 per cent instead of 0.7 per cent). This is collected and stored for subsequent use in fast reactors.

(b) The gas centrifuge

The separation of uranium isotopes using high-speed gas centrifuges again depends on the difference in the masses of the gas molecules. The technology of such centrifuges is relatively new, but it is expected that the process will require only about five to ten per cent of the power that the gas diffusion process uses.

A gas centrifuge is rather like a spin-dryer. As the gas is passed into the spinning centrifuge, the lighter molecules tend to stay close to the axis while the heavier ones tend to move out towards the periphery, thus causing a certain degree of enrichment to take place. To achieve useful output and enrichment levels, a large number of centrifuges have to be operated in cascades in the same manner as diffusion plant stages. Again, the depleted uranium is eventually stored for future use in fast reactors.

The world's first uranium enrichment plant was built at Oak Ridge in Tennessee during the Second World War, and it was here that the enriched uranium for the Hiroshima bomb was produced by means of the gaseous diffusion process. Since then, six further plants have been built – two in the USA, two in Russia, one in France, and one in Britain. Figure 3.2 shows an aerial photograph of Britain's enrichment facility – at Capenhurst in Cheshire – and gives some idea of the enormous size of such plants. Both the gaseous diffusion and the gas centrifuge processes are used at Capenhurst, the former taking place in the long row of buildings between the cooling towers and the road that runs from left to right through the centre of the plant, and the latter in the buildings on the other side of the road.

Manufacture of fuel elements

The metallic fuel for the British Magnox reactors is manufactured direct from purified natural uranium billets, which are remelted, alloyed, cast into rods and, finally, machined to size. The finished rods are then placed in cans made of the special magnesium alloy (magnox) which gives the reactors their name. To improve the heat transfer characteristics of the fuel elements, the surface area of these cans is extended by means of a series of fins.

Figure 3.3 shows part of the Magnox fuel assembly line at the Springfields plant of British Nuclear Fuels Ltd. Note that, at this stage of their careers, the fuel elements are only slightly radioactive, and can be

Figure 3.2 *The British uranium enrichment plant at Capenhurst in Cheshire*

handled by the process workers without any of the elaborate screening and safety precautions that are needed once they have been irradiated in a reactor.

Most other thermal reactors use ceramic fuel in the form of uranium dioxide, which is either manufactured from natural uranium (eg for the Canadian CANDU reactors) or from enriched uranium. In either case, the oxide is initially produced as a powder, and this is then formed into pellets using a hydraulic press. Next, the pellets are heat treated in order to convert them into ceramic form, and the pellets are then loaded into cladding tubes; these are filled with helium gas (to improve heat transfer between the fuel and the cladding) and sealed. Finally, the individual tubes are assembled into fuel elements using locating and support plates.

Irradiation in nuclear power stations

It is at this stage in the fuel cycle that the uranium undergoes fission in nuclear reactors. New fuel elements are transported to nuclear power stations in ordinary cases like any other cargo, and are loaded into the reactors to replace old fuel elements. Some Magnox, CANDU and AGR cores can be refuelled while the reactor is still producing power, a type of refuelling pattern that is described as *on load*. Most other reactor systems have to be refuelled *off load* (ie when the reactor is shut down).

On arrival at a power station, each new fuel element is first carefully inspected before it is loaded into the core. The loading is performed using a large, lead-shielded flask called a *charging machine*. Once in the reactor core, the fuel elements become progressively more radioactive owing to the steady bombardment with neutrons and the gradual build up of fission products (which are the most radioactive materials that it is possible to

Figure 3.3 *Part of the Magnox fuel assembly line at Springfields in England*

produce). Thus, elaborate precautions must be taken to protect the operating personnel when an irradiated element is eventually removed from the reactor. In practically all power stations, the charging machine serves a dual role as the unloading machine.

Storage

An irradiated fuel element is still generating a large amount of heat energy when it is removed from the core of a reactor. This is due to the fact that the highly active fission products release energy as they undergo radioactive decay. The spent fuel elements are therefore placed in lead-shielded flasks which are transferred to underwater storage areas (ponds), where they are left for several months in order to allow the worst of the radioactivity to die away. In fact, spent fuel elements lose up to 99 per cent of their activity in the first year of storage after being removed from a reactor.

It is here that the once-through fuel cycle ends. The spent fuel elements are either left in storage ponds close to the reactor, or, after a period of years, moved to a central storage facility from where they may eventually be removed for permanent disposal.

The reprocessing cycle

The alternative to simply storing the spent fuel elements is to reprocess them in order to remove the dangerous fission products and recover the unused uranium and valuable plutonium that they contain (see Figure 3.4).

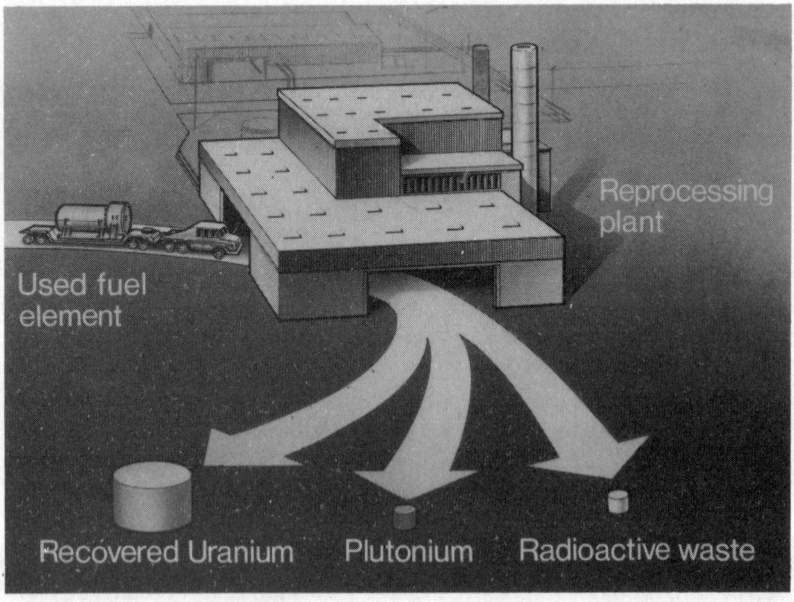

Figure 3.4 *The separation of the various components of spent fuel*

The simplest fuel cycle in which this type of separation occurs but which does not actually recycle the recovered plutonium is usually referred to as the *reprocessing cycle*. Here, the recovered uranium is sent back to the enrichment plant for use in the manufacture of new fuel elements, and the plutonium and fission products are sent to storage. Figure 3.5 shows the cycle in schematic form, and, by comparing it with Figure 3.1, we see that the reprocessing cycle and once-through cycle have many features in common. We shall therefore describe only those features that are new.

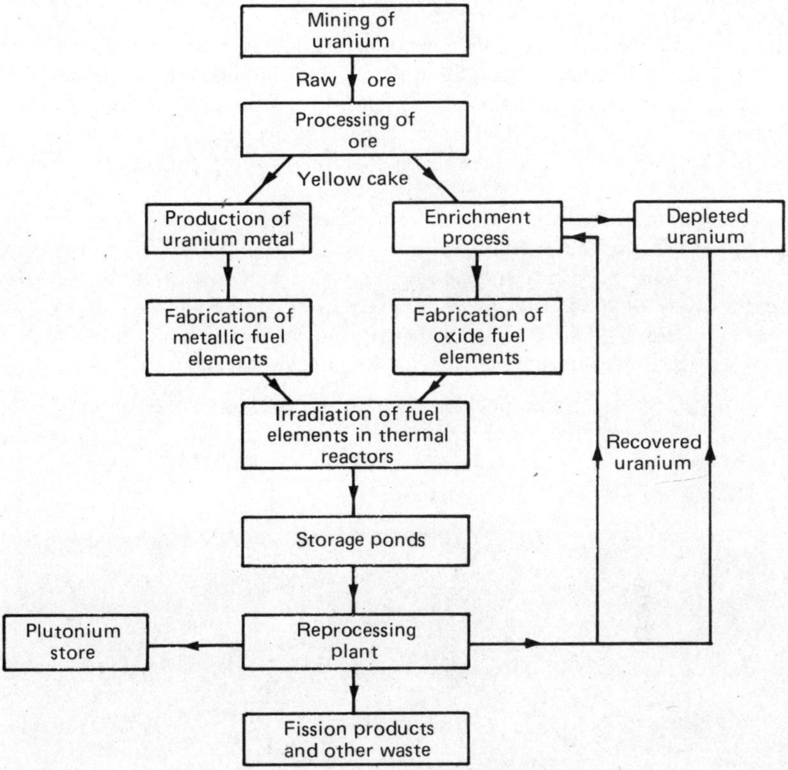

Figure 3.5 *The reprocessing cycle*

Storage

The spent fuel elements are first transferred from the reactor core to an underwater storage area, where they remain for a period of between one and ten years, depending on the type of fuel element. This allows their activity to fall to a level at which they can be safely moved and transported to the reprocessing plant.

Transport

Both road and rail transport are normally used to move the spent fuel elements from the storage ponds at reactor sites to the reprocessing plant. The still highly radioactive fuel elements are loaded into *transport flasks* – massive lead-shielded containers weighing between 50 and 100 tonnes and capable of carrying several tonnes of spent fuel. The flasks are usually transported by road from the power station to a suitable rail head, and hence to the reprocessing plant by rail.

Reprocessing

By the time the spent fuel elements reach the reprocessing plant, the most dangerous fission products (Xenon 133 and iodine 131) have decayed to very low levels, thus making chemical reprocessing technically and economically feasible. In Britain, such reprocessing is carried out at Windscale in Cumbria, an aerial photograph of the plant being shown in Figure 3.6.

At the moment, only the metallic fuel from Britain's Magnox programme is being reprocessed at Windscale. (A thermal oxide reprocessing plant (THORP) is currently under construction, and it is hoped that reprocessing of oxide fuel from the AGR programme will have started by the late 1980s.) The following description is therefore confined to the reprocessing of Magnox fuel. This involves the following main stages:

(a) dismembering the fuel element in order to expose the contents – uranium, plutonium and fission products;
(b) dissolving the contents in concentrated nitric acid;

Figure 3.6 *The fuel reprocessing plant of British Nuclear Fuels Ltd at Windscale in Cumbria*

(c) separating the uranium and plutonium from the fission products and from each other by chemical processes;
(d) dealing with the various solid, liquid and gaseous wastes that arise from these processes.

Because of the highly radioactive nature of the spent fuel elements, all these stages have to be carried out in heavily shielded rooms known as *cells* and *caves* using remote handling or automatically controlled equipment.

(a) Dismembering the fuel elements
Irrespective of the type of fuel element, it is necessary to remove the outer casing before the chemical reprocessing can begin. In the case of Magnox fuel, the element is first fed into a remotely operated machine which splits the can away from the actual fuel. Thus, only the fuel enters the dissolving stage of the process. (When oxide fuel elements are being reprocessed, the entire element is cut into thin slices using a large shearing machine, and the slices, which consist of both cladding material and fuel, then go on to the next stage.)

(b) Dissolving the spent fuel
This is usually done as a batch process. The chopped pieces of fuel element are put into a perforated stainless steel basket, which is then placed in concentrated nitric acid for several hours. The acid dissolves out the uranium, plutonium and fission products, leaving any solid wastes in the basket. The acid solution is then used as the plant feed for the separation process.

(c) The separation process
The process whereby the uranium and plutonium are separated from the fission products, separated from each other, and then purified is known technically as *liquid-liquid solvent extraction.* It is a well-established industrial process, and is very similar to that used in the mining industry to separate metals from their ores. The main difference is that, in this case, the various components to be separated out are all radioactive.

In the first stage of the process, the fission products, which represent up to three per cent of the contents of a spent fuel element, are separated out in an aqueous solution, which is then concentrated and sent to special stores for eventual disposal. The uranium (~97 per cent) and plutonium (~one per cent) are now free of the most dangerous radioactive components of the spent fuel and are themselves separated and purified.

The uranium that is recovered by reprocessing fuel from enriched uranium reactors still has a concentration of U235 that makes recycling to the enrichment plant an economic necessity. (This is the feedback route shown on the right-hand side of Figure 3.5.) That recovered from natural uranium reactors is generally so depleted in U235 that it is not worth recycling. Thus, it is simply stockpiled for eventual use as a breeder material in fast reactors. By using the reprocessing cycle instead of the once-through cycle, it is estimated that the raw uranium requirements of most thermal reactor systems could be reduced by up to 23 per cent.

The plutonium that is recovered from the reprocessing cycle is stockpiled for future use in manufacturing fast reactor fuel.

(d) Waste management

The final stage in the reprocessing cycle is the treatment of the various wastes that arise. These can be divided into two categories, namely, those that can be safely disposed of immediately, and those which must be contained and isolated from man for substantial periods of time (see Figure 3.7).

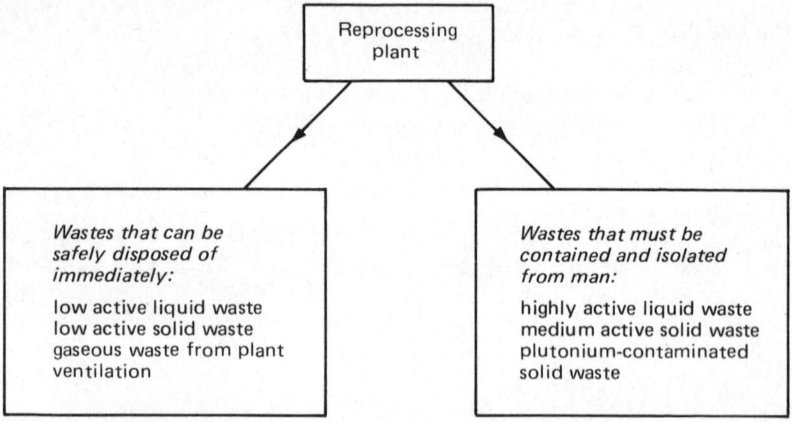

Figure 3.7 *Radioactive wastes from reprocessing operations*

Wastes that can be safely disposed of immediately

The *low active liquid and solid wastes* come from the process where the uranium and plutonium are separated from each other, purified and converted into solid form. The low active liquid waste is often discharged into the sea in a controlled manner and within strictly authorized limits. All such discharges are continuously monitored, and regular investigations are made of the take-up of radioactive waste products by fish, lobsters, milk, etc – the main pathways back to man.

The low active solid waste is usually buried underground, within the reprocessing plant, and represents no hazard either to the workforce or to the general public.

The *gaseous wastes*, which are released from some of the scrubbing and filtering processes, are produced only in very small quantities. Normally, these are discharged into the atmosphere through specially designed ventilation networks which ensure that they do no harm to the environment.

Wastes that must be isolated and contained
The *highly active liquid waste* consists of the separated fission products dissolved in nitric acid. In the UK, this is at present stored underground in large stainless steel tanks. Extensive feasibility studies have shown that highly active liquid waste can be dramatically reduced in volume and processed into a glass. This process, known as *vitrification*, is being developed by several countries, including Britain and France. The first industrial-scale vitrification plant is nearing completion at Marcoule in France, and it is expected that this will be followed by further plants in France and Britain towards the end of the 1980s. By the end of the century, it is expected that vitrified waste from the French and British nuclear power programmes will be sealed in lead-covered stainless steel containers so that it can be permanently (and safely) disposed of, deep down in stable geological formations.

The second type of waste that has to be isolated from man is the *medium active solid waste*. This consists of the remains of the fuel element wrappers and is β/γ active (see Chapter 4). The waste is usually packaged and stored, and is eventually likely to be disposed of in the same way as the glassified highly active waste (ie stored underground).

The *plutonium-contaminated solid waste* is low in β/γ activity but has significant α activity (again see Chapter 4), and consists mainly of the remains of the fuel pins after their contents have been dissolved out. Ultimately, it is expected that virtually all traces of the plutonium will be recovered from this waste, and that the remains will be stored prior to dumping at sea in internationally designated deep-water dumping areas.

The thermal recycle fuel cycle

In a variation of the reprocessing fuel cycle known as the *thermal recycle fuel cycle*, the recovered plutonium is fed back into the thermal reactor programme as a fuel. Although the plutonium is no more efficient than uranium as a thermal reactor fuel (rather less so, in fact), it has been estimated that using it in this way could save a further 16 per cent on raw uranium ore requirements. Thus, if the supply of uranium becomes scarce and the cost of the ore increases steeply, the thermal recycle fuel cycle may have some economic advantage. However, many people feel that there is a much more efficient way in which the plutonium produced from the thermal reactor could be used, namely, as a fuel for fast reactors. As was explained in Chapter 2, this would enable up to 60 per cent of the energy contained in natural uranium to be put to effective use compared with only one to two per cent in the use of a purely thermal programme.

The fast reactor fuel cycle

Fast reactor systems have yet to be used in commercial nuclear power stations, and so a true fast reactor fuel cycle will obviously take some considerable time to become properly established. It is expected that this will be done by gradually introducing fast reactors into commercial service alongside existing thermal reactors in what is known as a *fast reactor launch cycle* (see Chapter 9). Such a cycle is shown in schematic form in Figure 3.8, and we will now describe those features that have not been covered in previous sections.

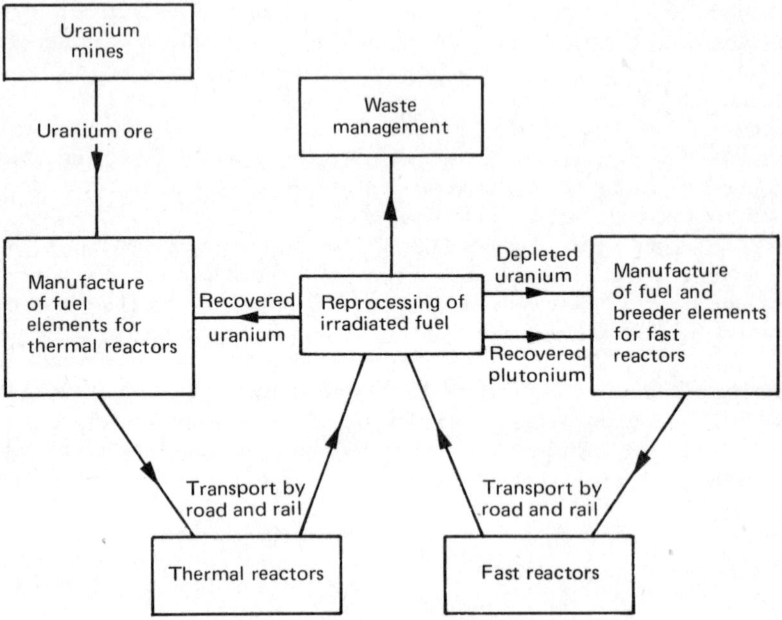

Figure 3.8 *The fast reactor launch cycle*

A fast reactor can either be fuelled with a mixture of plutonium and uranium or with plutonium alone. In either case, the fuel will probably consist of ceramic oxide pellets similar to those used in oxide-fuelled thermal reactors. The power-producing core of such a reactor will be surrounded by a blanket of breeder material – probably natural or depleted uranium, again in the form of ceramic pellets.

In a fast reactor launch cycle, both the plutonium and the uranium will have been recovered from a thermal reactor programme. Both materials are first prepared as powdered oxides, and are then formed into ceramic pellets by a similar process to that used to manufacture oxide fuel for thermal reactors. There is one important difference, however, since plutonium is *extremely* dangerous if ingested or inhaled. Stringent precautions must therefore be taken to ensure that no process worker ever comes in direct contact with plutonium, and that no plutonium ever escapes into the environment. At Windscale, specially designed glove boxes are used to isolate the plutonium in the fuel from the process workers.

As was stated in Chapter 2, a fast reactor can, by suitable core design, be made to operate either as a net consumer or as a net producer of plutonium. If the latter mode of operation is chosen, the fast reactor can, within a period of about 30 years, produce enough plutonium to provide fuel for a replacement for itself (a number of different designs of fast reactors have been proposed for which this 'doubling' time ranges from 18 years to 50 years). With a thermal reactor programme working in parallel with the fast reactor programme, a series of self-sustaining fast reactors could gradually be built over several decades. This would allow an established *fast reactor fuel cycle* to be introduced once the required number of fast reactors were in operation. What this means is that the right-hand side of Figure 3.8 could eventually operate quite independently of the left-hand side, although, for as long as thermal reactors were still in operation, supplies of depleted uranium and recovered plutonium would still be fed into the cycle.

As was said at the beginning of this chapter, a large number of different nuclear fuel cycles are possible. What we have tried to do is to describe the principal features of the main types. In Canada, the once-through cycle is used with that country's CANDU reactors, and the current moratorium on fuel reprocessing in the USA means that they too have reverted to this particular fuel cycle. In Britain and France, on the other hand, the fast reactor launch cycle is currently being built up.

We will return to the subject of nuclear fuel cycles in Chapter 9, when the possible long-term development of the world's various nuclear power programmes will be examined in detail, and also in Part IV, when the environmental and political implications of the various cycles will be discussed.

CHAPTER FOUR

Radioactivity – Its Nature and Biological Effects

The feature of the nuclear power industry that causes most concern to the general public is that many of the materials that have to be processed, stored and transported are radioactive. This final chapter of Part I describes what radioactivity is, and explains why radioactive materials are so dangerous if not properly handled.

Introduction

Radioactivity and ionizing radiations are not new phenomena. When the earth was formed some 4600 million years ago, it must have been much more radioactive than it is today because of the fall in activity that has taken place in the interim. It is believed, for instance, that most of the lead currently present in the earth's surface is the result of the radioactive decay of a whole series of heavier elements over this enormous length of time. In fact, a very crude estimate of the age of the earth can be obtained by assuming that all the lead found in association with uranium came originally from that uranium. The proportion of lead to uranium bears a close relation to the age of the rock, and therefore to the age of the earth itself.

The damage that large quantities of ionizing radiation can do to living cells only came to light during the early decades of the 20th century. In 1898, X-rays were discovered by Röntgen, and, at almost the same time, Becquerel discovered the radioactive properties of uranium. It was soon found that mice exposed to large doses of X-rays died shortly afterwards, and also that human skin became ulcerated and eventually developed cancerous growths when exposed to similar doses. Somewhat later, evidence was found that excessive radiation exposure could cause malignant changes in the bone marrow. By this time, some of the early radiation workers had already died from *leukaemia* (malignant bone marrow disease) and *aplastic anaemia* (destruction of the blood-forming cells). Unfortunately, these early workers had no means of measuring or even estimating their exposure to radiation, since it was not until the 1930s that the necessary techniques were developed.

The nature of radioactivity

A radioactive material is one whose nuclei are unstable and gradually *decay* to a more stable form by giving off a particle of some sort or by emitting energy. Since radioactivity is a nuclear phenomenon, the emission of a *particle* during radioactive decay causes the nucleus to change and a new element to be formed. Uranium 238, for example, decays to thorium 234 when it undergoes such decay.

The strength of a given radioactive source is measured in terms of its *activity*. The unit of activity is called the *curie*,* and a source having an activity of one curie emits 37 thousand million (3.7×10^{10}) particles per second. This is the number of particles emitted per second by one gramme of radium (the radioactive material first studied by Madame Curie – hence the name of the unit). In practice, however, the curie is an extremely large unit, most radioactive sources having activities that are more conveniently measured in *millicuries* (thousandths of a curie) or *microcuries* (millionths of a curie).

Another important characteristic of radioactive materials is their so-called *half-life*. This is the time that it takes for half the nuclei in a given sample of the material to undergo decay. We cannot identify which nuclei, or even which half of the sample, will decay. All that can be said is that, after a given interval of time (the half-life), half of the total number of nuclei originally present will have decayed. The half-life is therefore a purely statistical concept. Individual nuclei will have a range of different lifetimes, but, when a sufficiently large number are collected together, the decay behaviour of the group can be accurately predicted; the greater the size of the group, the more accurate the prediction. (Insurance companies use similar statistical methods when they are assessing life risks. For example, they can predict extremely accurately how many males in a country's total population will die within the next year, but they cannot say which ones they will be!) Figure 4.1 explains how the half-life concept works; it is rather like trying to reach a wall by halving the distance between you and the wall every time you take a step – you never quite get there!

The half-life is in fact a measure of the rate at which the strength of a radioactive source falls off with time, since the activity of a given source at any particular time is directly proportional to the number of undecayed nuclei remaining at that time. The half-life is also a measure of the intensity of a radioactive emitter, those with very short half-lives being very highly radioactive because of the fact that the great majority of their nuclei undergo decay over a comparatively short period. The half-lives of radioactive materials cover an enormous range – from thousands of millions of years for some naturally occurring emitters (such as uranium and thorium) to fractions of a second for some of the shorter-lived fission products.

* The curie is in the process of being replaced by a new unit – the *becquerel* – this being the activity of a source that undergoes one disintegration per second.

Number of half-lives that have elapsed	Number of original nuclei remaining	Number of nuclei that have decayed
0	N	0
1	N/2	N/2
2	N/4	3N/4
3	N/8	7N/8
4	N/16	15N/16
5	N/32	31N/32
.	.	.
.	.	.
.	.	.

Figure 4.1

If the same results are plotted on a graph, the resulting curve looks like that in Figure 4.2.

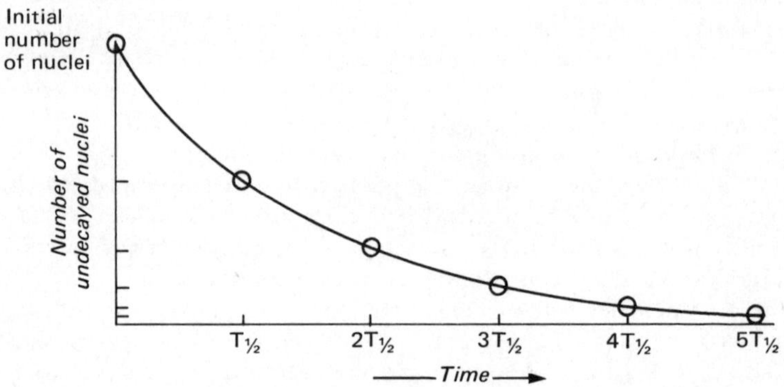

Figure 4.2 *A graph of the number of undecayed nuclei against time for a radioactive material*

The different types of radioactive emitter

There are three different types of radioactive emitter, known (for historical reasons as *alpha* (α) *emitters, beta* (β) *emitters* and *gamma* (γ) *emitters.*

Alpha emitters All elements higher than lead in the periodic table are intrinsically unstable, and tend to change to lighter elements by emitting an *alpha particle* from their nucleus. This alpha particle is in fact a helium nucleus, containing two protons and two neutrons – one of the most stable compound particles known. An example of an alpha emitter is uranium 238. This changes to thorium 234 as a result of the decay process, which can be represented as follows:

$$\text{U 238}\begin{matrix}\text{(92 protons)}\\ \text{(+ 146 protons)}\end{matrix} \rightarrow \text{Th 234}\begin{matrix}\text{(90 protons)}\\ \text{(+ 144 neutrons)}\end{matrix} + \alpha\begin{matrix}\text{(2 protons)}\\ \text{(+ 2 neutrons)}\end{matrix}$$

Beta emitters If an element has too many neutrons in its nucleus, it is unstable, and tends to change to a more stable configuration by changing one of its neutrons into a proton and emitting an electron in the process. This process is known as *beta decay*, and is very common in the case of fission products, virtually all of which are formed with far too many neutrons to be stable. An example of a beta emitter is carbon 14. This changes to nitrogen 14 as a result of the decay process, which can be represented as follows:

$$\text{C 14}\begin{matrix}\text{(6 protons)}\\ \text{(+ 8 neutrons)}\end{matrix} \rightarrow \text{N 14}\begin{matrix}\text{(7 protons)}\\ \text{(+ 7 neutrons)}\end{matrix} + \beta^{-}\ \text{(electron)}$$

Gamma emitters This is the simplest form of radioactive decay, and occurs whenever a nucleus is in an unstable, excited state and tends to change to a more stable, less energetic state by getting rid of its excess energy in the form of a *gamma ray* – a 'parcel' of electromagnetic radiation of very short wavelength. Gamma decay leaves a nucleus unchanged in basic character (ie with the same number of protons and neutrons), but in a more stable configuration. Gamma rays are in fact identical to X-rays of the same wavelength, differing from the latter only in their source of origin; γ rays come from *inside* the nucleus of the atom whereas X-rays come from *outside* the nucleus.

In practice, gamma emission is very often associated with beta emission, an unstable nucleus frequently undergoing both processes simultaneously, or, in some cases, undergoing gamma decay some time after it has undergone beta decay (because the resulting nucleus was still in an unstable, excited state).

Radioisotopes

We saw in Chapter 1 that elements can exist in different forms known as *isotopes*, the various isotopes of a given element having the same number of protons in their nuclei but different numbers of neutrons. In practically all cases, the naturally occurring isotopes of an element are stable, ie non-radioactive. It is, however, possible to produce unstable, radioactive forms of all the elements by artificial methods, for example by irradiating them with neutrons in the core of a nuclear reactor or isolating them from the fission products that are found in spent reactor fuel. These are known as *radioisotopes.* Radioisotopes can be α emitters, β emitters or γ emitters, although the last two are by far the most common.

Radioisotopes have found numerous applications in medicine, agriculture and industry, and their use is undoubtedly saving mankind enormous amounts of money each year. Figure 4.3 lists just a few of the most important areas in which they are making a major contribution.

Medicine
- diagnosis and treatment of certain diseases (eg cancer)
- study of biological processes within the human body
- study of disease-carrying organisms

Agriculture
- breeding of new varieties of plants
- studying effects of fertilizers
- disease control in plants and livestock
- animal nutrition

Industry
- fault detection
- quantity control
- diagnostic analysis
- tracer techniques
- preservation of food

Figure 4.3 *Some important applications of radioisotopes*

Figures 4.4 and 4.5 show two typical uses of radioisotopes. In Figure 4.4 we see a 7000-curie cobalt 60 source (a powerful gamma emitter) being used to irradiate a tumour located deep inside the body of a cancer patient. The machine is controlled by computer, and, by varying the direction from which the narrow beam of gamma rays is directed at his body, it is possible to ensure that only the tumour receives enough radiation to harm its cells – not the healthy tissue that surrounds it.

In Figure 4.5, we see maintenance engineers using a small iridium 192 source (another gamma emitter) to test parts of a Rolls-Royce RB211 aero engine for possible fatigue cracks. The process is similar to that by which cracks or breaks in human bones are detected using X-rays.

How ionizing radiation interacts with matter

Alpha, beta and gamma rays, together with X-rays and beams of neutrons of the type that are emitted from the cores of nuclear reactors, are collectively known as *ionizing radiation* because of the way in which they interact with matter. They all knock electrons out of the atoms through which they pass (ie *ionize* them). The physical effects of radiation are best measured in terms of the amount of ionization that the radiation produces or the amount of energy that it causes to be released (the two are equivalent). Since energy is used up in ionizing atoms, the ionizing radiation becomes progressively weaker as it passes through matter until, eventually, it has insufficient energy to 'knock out' any more electrons. When this stage is reached, the radiation can be said to have been *stopped*. Because of their different natures and energies, the different types of ionizing radiation have radically different penetrating powers.

Although it is relatively heavy and can be emitted with a very high velocity (up to 1/10 of the speed of light), the α particle has a low penetrating power because it loses its energy very quickly as it passes through matter. A beam of α particles is easily stopped by a sheet of paper, a protective rubber glove or a thin sheet of aluminium foil. Even a few centimetres of air will stop α particles, and α radiation only causes serious damage to the body when emitted internally (as would be the case if an α-emitting material were swallowed).

Electrons (in the form of β radiation) cause ionization in much the same way as α rays, but do not lose their energy so quickly. Thus, β rays are much more penetrating than α rays, having a range of about one metre in air. A few millimetres of lead will stop all the β particles from a source, however. Like α radiation, β radiation can cause damage to thc skin and to organs in the body when emitted internally.

γ radiation consists entirely of electromagnetic energy and therefore has no mass or charge. Because of this, the effective range of γ rays in a given material is very much greater than that of either α particles or β radiation. In air, for example, the effective range of γ rays is several kilometres. The intensity of γ radiation can be reduced to any desired level by using an appropriate thickness of shielding material. A thickness of seven centimetres of lead, for example, reduces the intensity of a beam of γ rays to roughly 1/100 of its original value, while a thickness of 14 cm

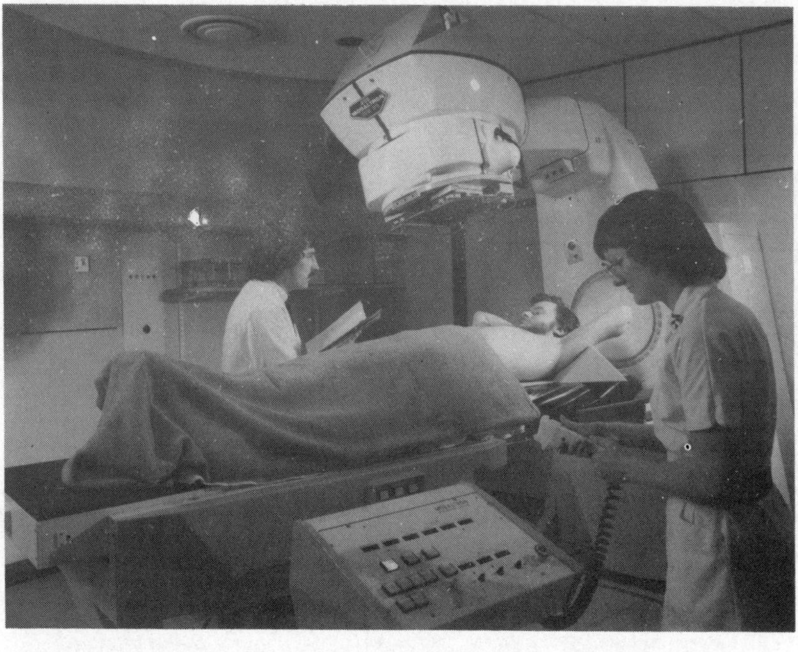

Figure 4.4 *A cancer patient receiving gamma rays from a cobalt 60 source*

reduces the intensity to 1/10,000 of its original value, a thickness of 21 cm to 1/1,000,000 and so on. X-rays behave in a similar manner.

Since they carry no electric charge, neutrons can only produce ionization in an indirect manner. They do this by colliding with the nuclei of the atoms of the material through which they pass, transferring some of their energy to these nuclei in the process. Thus, it is these 'recoil nuclei' that are responsible for producing the ionization, not the neutrons themselves. Note that neutrons are not emitted by radioactive materials in the same way as α, β and γ rays, the main sources of these particles being nuclear reactors and atomic explosions.

The effects of ionizing radiation on living cells

All living material is composed of basic structural units known as cells, generally less than 1/1000 centimetre across. These cells are generally irregular in shape, and consist basically of a small central body known as the nucleus embedded in a liquid known as *cytoplasm*. The cytoplasm of one cell is separated from that of neighbouring cells by the membranes that form the cell walls. A schematic diagram of a cell is shown in Figure 4.6.

The nucleus of a cell contains a number of string-like bodies known as *chromosomes* arranged in pairs, and these, in turn, contain the *genes* that carry the complete genetic plan of the living creature to which the cell

Figure 4.5 *Using gamma rays from an iridium 192 source to test parts of an aero engine for possible cracks*

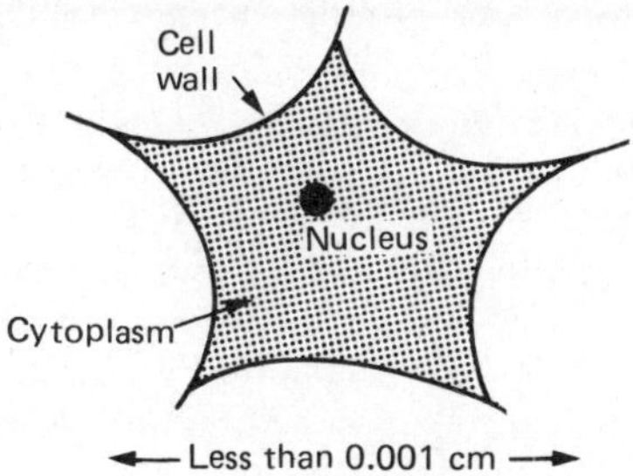

Figure 4.6 *A schematic representation of a living cell*

belongs.

When living matter is bombarded with ionizing radiation, that radiation which passes through the cytoplasm of a cell has little or no effect on its activity. Any radiation that passes through the nucleus of a cell, however, may cause considerable damage to the cell – either by impairing its ability to carry out its specific bodily function or by impairing its ability to reproduce. Practically all the cells in a creature have a limited life, and must therefore continuously reproduce themselves if the creature is to survive. The damage to a cell may be caused in two ways. Firstly, the ionizing radiation may strike an individual chromosome, and may either completely destroy it or alter the structure of one or more of the genes. Alternatively, the ionizing radiation may pass through the nucleus without actually striking any particular chromosome, but, as it does so, may split up some of the water molecules in the fluid surrounding the chromosomes into highly active H^+ and OH^- ions, which may then damage the chromosome by reacting with their genes. Any alteration to the chromosomes of a cell that results from either of these two mechanisms may prevent the cell from reproducing, or may cause it to produce cells that do not carry out their proper function. Thus, in either case, the cell has effectively been killed. Ionizing radiation can also cause temporary damage to cells, ie damage that is eventually repaired by internal 'healing' mechanisms.

In order to quantify the physical and biological effects of ionizing radiation on living matter, we need to be able to measure both the amount of the radiation and its effectiveness. The *physical effects* are measured in terms of the amount of ionization that the radiation produces, or the amount of energy that it causes to be dissipated (the two are equivalent). The standard unit is the *rad*,* a dose of one rad being the amount of

* Both the *rad* and the *rem* are in the process of being replaced by new units in order to bring the units of absorbed dose and dose equivalent into the SI system (the internationally agreed system of units used in science). The rad is being replaced by the *gray*, where one gray ≡ 100 rads, and the rem by the *sievert*, where one sievert ≡ 100 rems. We have decided to use the old units in this book, however, since it will be some time before the new units gain general use. Also, practically all the literature referred to in the Bibliography uses the rad and the rem.

radiation that causes 1/100 joule of energy to be dissipated in every kilogramme of matter on which it falls (the joule is defined on p 19 in Chapter 1).

When measuring the *biological effects* of ionizing radiation, it is necessary to take account of the fact that some types of radiation have a greater effect (dose for dose) than others. A dose of one rad of fast neutrons, for example, produces ten times more biological damage than one rad of γ rays. This difference is taken into account by multiplying the absorbed dose for each type of radiation by a so-called *quality factor* which reflects the ability of the particular type of radiation to cause damage. The relative quality factors for the different types of ionizing radiation are shown in Figure 4.7.

Radiation	Quality factor
Electrons, X-rays, γ rays	1
Thermal neutrons	3–4
Fast neutrons	10
Protons	10
α particles, nuclear fragments	20

Figure 4.7 *The relative biological effectiveness of different types of radiation*

The resulting quantity is known as the *dose equivalent*, the unit of which is the *rem*. Since the rem is an extremely large unit (roughly six times the total amount of radiation that the average British citizen receives from all sources, both natural and man-made, in the course of a year) a smaller unit known as the *millirem* (equal to 1/1000 rem) is commonly used in practice.

The effects of whole-body radiation doses on human beings

When a human being is subjected to a whole-body dose of ionizing radiation, the effect depends strongly on the size of the dose, as shown in Figure 4.8.

The effects of large localized doses

In the case of localized doses of radiation, the effects depend on the part of the body that is irradiated, some parts being much more sensitive than others. The most serious effects are mainly found in tissues where cells are continuously in division, the blood-forming organs in the bone marrow, the skin, the linings of the gut and the sperm-production cells. Non-dividing systems, like nerve and muscle cells, do not normally show any effects until very large doses of radiation are received.

Below 25 rems	no noticeable effects
25–100 rems	slight changes in blood count, but no other noticeable effects apart from possible chromosome damage
100–200 rems	in up to 50% of cases, vomiting within a few hours, followed by fatigue and loss of appetite which generally persists for several weeks; moderate damage to the blood-producing bone marrow cells, lasting considerably longer
200–300 rems	same short-term symptoms in up to 100% of cases, together with even more severe long-term blood damage; mortality in a few cases within several months
300–600 rems	in 100% of cases, vomiting, fatigue etc within 2 hours; loss of hair within 2 or 3 weeks; severe blood damage, accompanied by haemorrhage and increased sensitivity to infection; up to 80% mortality within 2 months
600–1000 rems	in 100% of cases, vomiting, fatigue etc within 1 hour, followed by loss of hair, severe blood damage, etc; practically 100% mortality within 2 months
Over 1000 rems	100% mortality within 2 months

Figure 4.8 *The effects of whole-body radiation doses on human beings*

Blood forming organs The red blood cells, the platelets and the granulocytes are made in the bone marrow, while the lymphocytes are mostly made in the spleen and lymph nodes. A dose in excess of 500 rems will drastically reduce the efficiency of these various organs, resulting in less resistance to infection, bleeding and severe anaemia.

Skin No noticeable effect is produced by a localized dose below about 500 rems, but, for doses above this, temporary reddening of the skin and loss of hair occur within a few days. Over about 1000 rems, the skin begins to peel, and, if the dose is above 3000 rems, local superficial damage known as *moist desquamation* is produced within a week. Here, the outer layer of the skin is lost, and an ulcer in the form of an open sore is produced; this generally takes several months to heal. For even higher doses (greater than 5000 rems), permanent destruction of the tissue (known as *necrosis*) occurs within a few weeks, resulting in an extremely painful open sore that is unlikely ever to heal. The changes in the skin closely resemble the effect of a thermal burn. The difference is that, with radiation, the damage may only become apparent after a week or so, whereas normal thermal burn damage (blisters etc) becomes apparent after a few minutes.

Intestines Loss of lining cells, due to radiation damage, can result in loss of fluid and electrolytes (such as sodium).

Testes According to the dose, the loss of cells which make sperm may be either temporary or permanent. Temporary sterilization can be induced by a single dose as low as ten rems, and permanent or prolonged sterility will result from doses above about 500 rems.

Ovaries Unlike the testes, the ovary is not continuously manufacturing new cells. At birth, it is provided with its entire stock of primitive cells which are capable of turning into ova. As these primitive cells cannot be replaced, their destruction leads to permanent sterility. A dose of about 250 to 300 rems to the ovaries in early reproductive life will cause sterility, but, towards the menopause, when fewer primitive cells remain, a smaller dose than this will sometimes result in sterility.

The embryo In its early life in the womb, many cell divisions have to take place, and the resulting new cells have to be organized into tissue and organs. During this period, the embryo is extremely sensitive to radiation, which can result in structural abnormalities and deformities. An investigative antenatal X-ray dose of two rems, for example, can increase the risk of early leukaemia, and a localized dose of 100 rems can induce an abortion.

Doses of over 50 rems to *any* organ of the body have an adverse effect on its function because of the proportion of cells that are killed by the radiation. This effect can, however, be used to the *benefit* of some cancer patients by making it possible to carry out X-ray or γ-ray therapy. Here, a very localized high dose is used to kill the cancerous cells in a tumour (see Figure 4.4).

Long-term effects of radiation

In addition to the specific effects described above, there are a number of other effects of exposure to radiation which only appear (if at all) after a period of several years or tens of years. Possibly the best known of these are the various forms of cancer, many of which (leukaemia, for instance) are thought to be capable of being produced by exposure to radiation – either in the form of a single dose or in a large number of small doses over a long period. Epidemiological studies on groups of people who have been irradiated have provided information which has enabled some quantitative estimates of the risk of delayed effects following irradiation to be made. (We shall be taking a closer look at some of this evidence in Part IV, but, for the moment, we shall simply summarize the main points.)

Data on the long-term effects of radiation are normally collected from groups of people who have been exposed to relatively large doses of radiation (eg survivors of the Hiroshima and Nagasaki bombs). These doses are greatly in excess of the internationally agreed permitted levels, and so it is necessary to extrapolate downwards to these much lower levels in order to estimate the long-term effects of radiation on the vast majority of the people in the world. As the evidence stands at the moment, it is not really possible to decide whether there is such a thing as a 'low threshold dose' below which no delayed effects occur, or whether all radiation exposure carries some penalty. If the former is in fact the case, the threshold dose will depend on the location of the irradiation and the ability of that part of the body to 'repair' itself. In some instances, the threshold value may be lower than the natural background dose received, in which case delayed effects from 'natural causes' will mask any other effects.

As has been described earlier, radiation can cause damage to the blood-forming cells in the bone marrow. It is thought that continuous exposure to relatively small doses of radiation can ultimately produce permanent damage to the bone marrow. If detected sufficiently early, this can generally be cured, but, if not, it can eventually lead to death if the overall dose is sufficiently high. It is also believed by some workers that exposure to radiation can hasten the ageing process.

Genetic effects of radiation

In addition to the potential hazard that it represents to individuals, ionizing radiation represents a potential hazard to the human race as a whole because of the harmful genetic effects it can produce. By damaging the genes in the chromosomes of the reproductive cells, such radiation can give rise to *mutations*, ie individuals with one or more characteristics that differ from those of the rest of the species to which they belong.

Every human being begins life as an individual cell formed by the fusion of two reproductive cells, one from each parent. The reproductive cells differ from the other cells in the body in that they contain only 23 chromosomes, as opposed to the normal 46. When two reproductive cells combine to form a fertilized egg, the latter contains the full complement of 46 chromosomes, arranged in 23 pairs, with one member of each pair coming from each parent. All the detailed physical characteristics of an individual are determined by the genes that make up the chromosomes, each characteristic being determined by one pair of genes – one of which comes from the mother and the other from the father. Let us take eye colour as an example. If the two genes which determine eye colour are the same (eg both tending to produce brown eyes), then the individual will definitely have brown eyes. If the two genes are different, however (eg one gene tends to produce brown eyes and the other blue eyes), then the colour of the individual's eyes will be determined by the gene which exerts the more powerful influence.

Genes can in fact be divided into two groups, namely *dominant* and *recessive*. A dominant gene is one that always asserts its influence over the particular characteristic that it is associated with, and a recessive gene is one that only asserts its influence if it happens to be paired with a similar gene. In the example described above, the 'brown-eyes' gene is the dominant one, so that a brown-eyed person can have either two 'brown-eyes' genes or one 'brown' and one 'blue'. A blue-eyed person, on the other hand, can only have two of the recessive 'blue-eyes' genes.

Ionizing radiation can alter the structure of the individual genes in a particular cell. If this structural alteration occurs in one of the genes in a reproductive cell, and this cell is used in the formation of a fertilized egg, one of two things can then happen. Firstly, the altered gene may prevent the egg from reproducing itself, thus effectively killing it. Alternatively, the altered or mutated gene may be accepted by the egg cell. If this is the case, the effect will depend on whether the gene is dominant or recessive. For a dominant mutated gene, the particular characteristic that it controls will appear in the individual that develops from the egg. Mutations that appear in this way are mostly fairly trivial, but at least 99 per cent are harmful to the individual concerned. If, on the other hand, the mutated gene is recessive, the mutation will not appear in the individual that develops from the egg, but may be handed down by that individual to future generations. If a male and female descendent who both carry the same recessive mutated gene subsequently mate, the mutation may appear in the offspring. Thus, a harmful mutation produced by ionizing radiation can either take effect immediately (resulting in the deformation or even the death of the individual produced) or can lie *dormant*, only appearing after several generations. Given a sufficiently long time, such harmful recessive genes can in fact spread through an entire population.

Mutations are not, however, a rare occurrence, and not all mutations are harmful. They occur naturally all the time in all species of plants and animals, and are believed by geneticists to be responsible for the evolution of species. The process of natural selection causes the harmful mutations to disappear (owing to the early death or infertility of the individual concerned), and enables the rare beneficial mutations to spread throughout the species until they become universal. Only about one per cent of all mutations are thought to be beneficial, and this probably explains why evolution is such a slow business.

It is thought that roughly 90 per cent of genetic mutations occur spontaneously (produced by thermal excitation of the gene molecules) and that the remaining ten per cent are produced by ionizing radiation. Until recently, all of this originated from natural sources such as radioactive rocks, cosmic rays and radioisotopes that occur naturally in the body. During the last 60 years or so, however, man has been steadily increasing the rate at which mutated genes are produced by radiation – eg by medical X-rays and fall-out from nuclear weapons. This will almost certainly have *some* genetic effect on future generations, although whether or not the effect will be noticeable is still a subject of debate – as is the question of whether the nuclear power industry will produce any noticeable genetic effect. We will resist the temptation to digress on this point, however, and will leave further discussion of the matter to Chapter 11, when it will be examined in detail.

PART TWO

Why We Need Nuclear Power

In 1920, over 90 per cent of the world's energy requirements were met from a single resource – coal. By 1979, in contrast, the contribution from coal had dropped to below 30 per cent and the most important overall source of energy had become oil. This change in usage pattern has not been uniform, however, for while coal still accounts for almost 50 per cent of the total energy input to the communist world, in the non-communist world the corresponding figure is only about 20 per cent.

In the following three chapters, which constitute Part II of the book, we will first take a detailed look at recent energy usage patterns in the non-communist world, and will then try to suggest ways in which future energy demands might be estimated, and, more importantly, met. In the course of this analysis, we will show how there is likely to be a severe energy shortage soon after the beginning of the 21st century unless significant use is made of nuclear power. To illustrate this argument, we will use the USA and Britain as case studies.

Note on energy data given in Part II

All primary energy figures are quoted in units of *millions of tonnes of coal equivalent* (abbreviated *mtce*). This has been done to make comparisons between primary energy inputs easier to understand.

In order to achieve this uniformity of presentation, it has been necessary to convert the various indigenous energy units (eg barrels of oil and cubic feet of natural gas) into our standardized units (mtce) by using appropriate conversion factors. The main sources of energy data (the United Nations, the major oil companies and national government departments of energy) all use different conversion factors, however, and, in some cases, do not even quote the primary energy figures in the same indigenous units. The production of graphs and figures, particularly for Chapter 5, has therefore involved a considerable amount of 'detective work'.

As a result of the different conventions used, there will inevitably be slight inconsistencies between the energy data given in this book and in other sources. The differences are, however, quite small, and amount to only a few per cent at worst.

The following approximate calorific values (based on those used in the publication *BP statistical review of the world oil industry 1979*) have been used for the main primary energy resources:

1 mtce = 1 million tonnes of coal
≡ 0.67 million tonnes of oil
≡ 0.78 thousand million cubic feet of natural gas.

In addition, it has been assumed that 2680 million units (kWh) of electricity are produced from one mtce in a modern power station.

Note on terminology

In Chapter 5, the terms 'communist world' and 'non-communist world' are used frequently. The first is intended to mean the USSR, Eastern Europe and China, while the second embraces the rest of the world. However, since the industrialized countries of the non-communist world consume over 80 per cent of that sector's energy, most of the discussion will inevitably relate to the consumption patterns of the industrialized non-communist countries.

Authors' note:
This book was written in the first half of 1981, before the 1980 energy figures became available and before the full severity of the current world recession was realized. The main arguments in Part II are, however, still valid.

CHAPTER FIVE

Use of Energy in the Non-communist World – the Changing Pattern since 1950

In this chapter, we will review the way in which the energy consumption pattern of the industrialized non-communist world has changed since 1950. This period is interesting in that it includes both the era of unprecedented economic expansion that resulted from the adoption of Keynesian economic policies throughout most of the non-communist world after the end of the Second World War and also the period in which this expansion was temporarily checked by the Middle East oil crisis of 1973-75.

After reviewing the world picture, the situations in the USA and in Britain will be examined in more detail. The USA has been chosen because it is by far the largest consumer of energy in the world today, and, as such, exerts considerable influence on the energy policies of all other countries. Britain has been chosen because it is a typical example of a non-communist industrial nation (and, of course, because the authors are British!).

Primary energy consumption in the world (1950–79)

Figure 5.1 shows the changes in primary energy consumption that have taken place in the world as a whole and in the non-communist world over the period 1950-79. Let us examine some of the main features of this graph, starting with the world as a whole.

The world

Over the period 1950-79, the rate of energy consumption more than trebled. In the decade to 1960, the average annual rise in energy consumption was 5.4 per cent, and, between 1960 and 1970, the figure was only slightly lower at 5.3 per cent per annum. This steady increase in demand for energy was temporarily halted in 1974-75, but, since then, it has continued to rise at about 3 per cent per annum. Even allowing for an increased demand being necessary because of increasing population, the world was consuming considerably more energy per capita in 1979 than it was in 1970. This is shown clearly in Figure 5.2 overleaf.

The communist block countries

The energy consumption pattern of the USSR, Eastern Europe and China

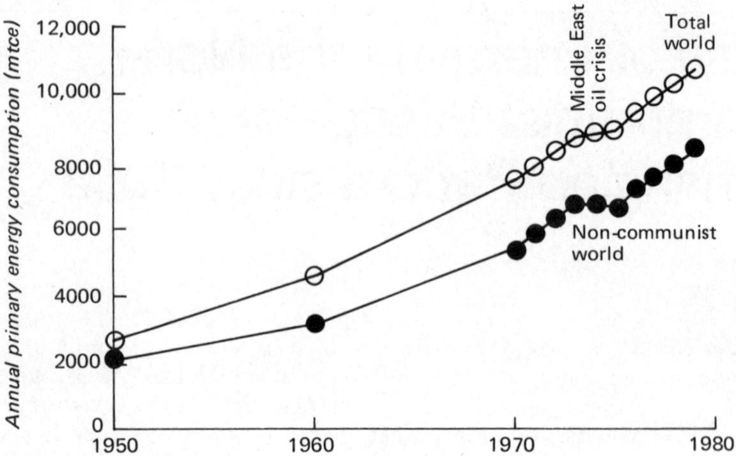

Figure 5.1 *Primary energy consumption (1950-79) – world and non-communist world*

	1970	1973	1974	1975	1976	1977	1979
World population (millions)	3610	3860	3890	3967	4044	4124	4300
Energy consumed per capita (tonnes coal equivalent)	2.15	2.30	2.30	2.26	2.34	2.37	2.43

(Population figures from UN Statistical Year Books. 1979 figures are estimates.)

Figure 5.2 *Per capita primary energy consumption (1970-79) – world*

is represented by the difference between the two lines plotted on the graph shown in Figure 5.1. As stated in the introduction to Part II, coal still supplies almost 50 per cent of their total energy demand, with oil contributing less than 30 per cent. It might have been expected, therefore, that these countries would be less affected by the Middle East oil crisis than the oil-dependent Western world, and the figures show that this was indeed the case. Between 1970 and 1979, the per capita energy consumption of the communist world increased steadily, almost totally unaffected by the traumatic events of 1974, rising from 1.83 tonnes of coal (well below the world average) to almost 2.5 tonnes (just above the world average). Indeed, between 1975 and 1979, the total energy consumption of the communist block countries increased at an average rate of 5.1 per cent per annum – well above the rate of increase in the West.

The non-communist world

Figure 5.3, which corresponds to the lower line of the graph in Figure 5.1, shows the changing pattern of energy consumption in the non-communist world over the period 1950-79.

Year	Coal	Oil	Natural gas	Hydropower	Nuclear power	Total
1950	1151	705	220	134	–	2210
1960	1212	1291	503	280	~1	3287
1970	1287	2908	1100	396	28	5719
1971	1230	3060	1170	413	40	5913
1972	1203	3264	1224	422	54	6167
1973	1246	3508	1243	429	68	6494
1974	1250	3389	1256	449	85	6429
1975	1229	3294	1201	465	120	6309
1976	1295	3519	1242	466	143	6665
1977	1321	3612	1242	481	178	6834
1978	1332	3721	1274	518	203	7048
1979	1431	3730	1292	529	209	7191

Figure 5.3 *Primary energy consumption (1950-79) – non-communist world (in mtce)*

Over this period, the rate of energy consumption more than trebled. In the decade to 1960, the annual rise in energy consumption was roughly four per cent, and, between 1960 and 1970, this rose to almost six per cent per annum. Total energy consumption actually fell slightly between 1973 and 1975, but, since that time, it has risen at roughly 3.3 per cent per annum. As the figures given in Figure 5.4 clearly show, the recent rise in energy consumption cannot be entirely attributed to the rising population, since per capita energy consumption has also shown a systematic increase.

Figure 5.3 shows how our dependence on oil has changed over the years, and also explains why we were relatively more affected by the events of 1974 than were the communist block countries. Figure 5.5 demonstrates this comparison even more dramatically.

In 1950, over half the non-communist world's total energy input was provided from coal. By 1979, however, the contribution from coal had dropped to below 20 per cent, with oil now providing nearly 52 per cent of the total input. This change from coal to oil as the major energy

	1970	1973	1974	1975	1976	1977	1979
Population (millions)	2492	2689	2707	2767	2827	2890	3000
Energy consumed per capita (tonnes coal equivalent)	2.29	2.42	2.37	2.28	2.36	2.36	2.40

(Population figures from UN Statistical Year Books. 1979 figures are estimates.)

Figure 5.4 *Per capita primary energy consumption (1970-79) — non-communist world*

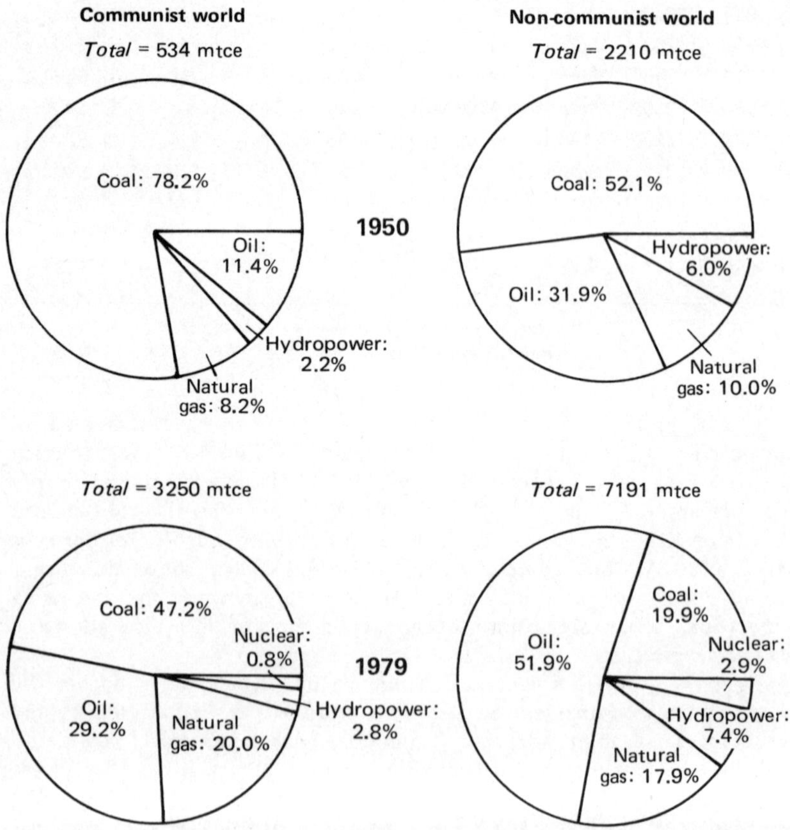

Figure 5.5 *Primary energy consumption by source — communist and non-communist world*

provider took place steadily between 1950 and 1970 – a period of considerable industrial expansion and relatively low oil prices. Since 1970, the drift away from coal has slowed down considerably as progressively higher oil prices have been introduced and nuclear power has begun to make a small but significant contribution to our total energy input. We are still, however, left with largely oil-based economies, although the high price of oil is now persuading us to look for alternatives.

In order to focus more clearly on the industrialized countries of the Western world, we will now take a closer look at the energy consumption patterns of two specific nations – the USA and Britain.

Primary energy consumption in the USA (1950-79)

In terms of energy consumption (as in many other ways) the USA is an extreme case, with six per cent of the world's population consuming a third of the world's energy. (It is also true, of course, that the USA *produces* roughly a third of the world's energy.) Since energy consumption cannot easily be separated from industrial prosperity and (hence) from economic wealth, it is unlikely that this situation will change dramatically over the short term. This does not mean that the USA can adopt an 'I'm all right, Jack' attitude, however, as we shall see in Chapters 6 and 7.

The primary energy consumption figures for the USA are given in Figure 5.6 and the changing pattern of resource usage is shown in Figure 5.7.

Year	Coal	Oil	Natural gas	Hydropower	Nuclear power	Total
1950	448	485	216	37	–	1186
1960	361	737	447	56	<1	1601
1970	475	1094	778	93	8	2448
1971	455	1133	793	101	14	2496
1972	474	1222	805	105	20	2626
1973	511	1289	803	107	31	2741
1974	506	1240	773	118	43	2680
1975	510	1215	711	115	65	2616
1976	548	1304	726	110	72	2760
1977	567	1372	700	86	94	2819
1978	565	1408	703	117	105	2898
1979	611	1367	695	117	100	2890

Figure 5.6 *Primary energy consumption (1950-79) – USA (in mtce)*

Perhaps the most striking feature of the USA's primary energy consumption pattern is its heavy reliance on oil and natural gas. In the past, this has almost entirely been sustained by the USA's indigenous reserves. In 1950,

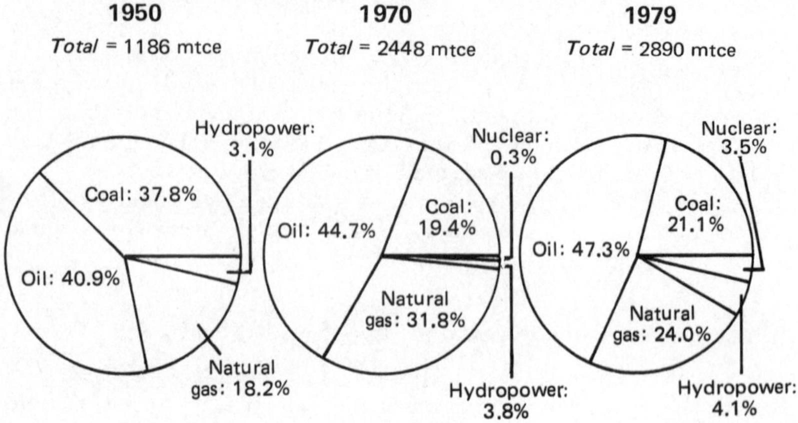

Figure 5.7 *Primary energy consumption by source (USA) – the changing pattern*

for instance, the USA consumed 63 per cent of that year's world oil production. By 1979, the figure had dropped to 29 per cent, but, during the same period, annual consumption of oil had almost trebled. In the case of natural gas, the 1950 share was 86 per cent, and, by 1979, 36 per cent, with the actual annual consumption again more than trebling. In 1950, almost 60 per cent of the USA's primary energy input came from oil and natural gas; by 1979 this figure had increased to over 71 per cent.

Coal remains an important energy resource in the USA, and its use during the 1960s increased steadily in order to cope with the growing demand for electricity. Since 1974, coal's market share has again increased by a small but significant amount. Indeed, with the recent slowing down in the development of nuclear power in the USA, coal may well continue to increase its share in the USA's energy market.

In the decade to 1960, the average annual rise in energy consumption was three per cent; between 1960 and 1970 it was 4.3 per cent per annum. In common with the rest of the non-communist world, total energy consumption actually fell between 1973 and 1975, but has risen at an average of 1.5 per cent per annum since then. In 1979, the figure fell slightly from the 1978 level; at the time of writing there is no indication as to whether or not this is a trend which will be maintained in future years, although, on the basis of past history, it seems highly unlikely that this will be the case.

Energy usage by sector

Figure 5.8 compares the energy usage patterns in the USA at three different times since the 1950s. The table shows a gradual emergence of transportation as a major energy consumer comparable with industry.

It also illustrates the comments made above about the USA's dependence on oil and natural gas. In 1979, for example, 59 per cent of all oil used in the USA was devoted to the transport sector and 52 per cent of all natural gas was consumed in the industrial sector. Use of coal, on the other hand, seems to be almost completely confined to industry and electricity generation.

	1962	1972	1979
Residential (domestic)	19.0%	17.8%	18.3%
Commercial and others	11.5%	13.2%	10.0%
Industrial	42.0%	38.3%	38.4%
Transport	27.5%	30.7%	33.3%

Figure 5.8 *Primary energy usage (USA) by sectors*

Generation and use of electricity

From the 1950s to 1973, the demand for electricity in the USA increased by an average seven per cent per annum. As a result of the dramatic increase in oil and coal prices since 1974, however, the rate of increase in demand from 1974-77 fell to only about four per cent per annum.

The USA's installed generating capacity increased steadily from 189 thousand megawatts (189 GW) in 1962 to 528 GW in 1977. By then, the peak demand was roughly 397 GW, giving a reserve margin of almost 33 per cent. This relatively high margin illustrates the way in which a major change in demand such as that produced by the 1974 Middle East oil crisis can upset the plans of the electricity supply industry. During the 1960s, when stations due to come on stream during the 1970s were ordered, it was expected that demand would continue to increase at the current rate of seven per cent per annum; thus, when the rate of increase in demand fell to four per cent after 1974, the industry found itself with more plant than it needed.

In 1978, roughly 31 per cent of the primary energy consumed in the USA was used in the generation of electricity. This amounted to roughly 900 mtce, the distribution among the various primary energy resources being shown in Figure 5.9.

In 1978, President Carter signed the various bills that constitute the National Energy Act – bills that will almost certainly have a profound effect on the way in which America's electricity is produced in future years. One bill prohibits the use of natural gas or oil for new power plants and also prohibits the use of natural gas by existing power plants after 1989. In addition, it limits the amount of natural gas that can be used in existing power plants during the remainder of the 1980s. The overall effect of this bill will probably be to increase the proportion of coal-generated

Primary fuel	Amount (mtce)	Per cent of total
Coal	417	46.1%
Oil	145	16.0%
Natural gas	119	13.2%
Hydropower and geothermal	118	13.1%
Nuclear power	105	11.6%

Figure 5.9 *Use of primary fuels in the generation of electricity (1978) – USA*

electricity, reduce the proportion of oil-generated electricity and, of course, gradually eliminate gas-fired power plants from the system.

The use of electricity is generally defined in terms of three principal sectors – residential (or domestic), commercial and industrial – as can be seen in the pie chart shown in Figure 5.10. Since the 1960s, the residential and commercial shares of the electricity market have gradually increased at the expense of the industrial sector, while the total demand has itself continued to increase.

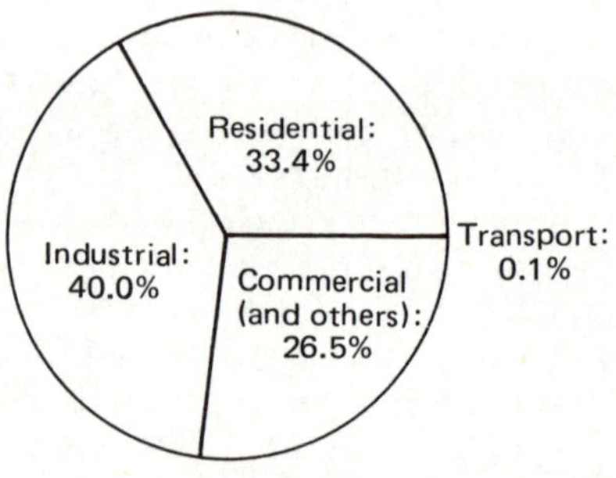

Figure 5.10 *Use of electricity – 1978 (USA)*

Energy consumption per capita

Perhaps not surprisingly, the energy consumption per capita in the USA is much higher than the average figures for the communist or non-communist worlds. The national trend since 1970 follows the by now familiar pattern, namely, a relentless gradual increase checked only by the events of 1974. Figure 5.11 shows the trend in summary form.

	1970	1973	1974	1975	1976	1977	1979
Population (millions)	204.9	210.4	211.9	213.6	215.1	216.8	220.2
Energy consumed per capita (tonnes of coal equivalent)	11.95	13.03	12.56	12.25	12.83	13.00	13.12

(Population figures from UN Statistical Year Books. 1979 figures are estimates.)

Figure 5.11 *Per capita primary energy consumption (1970-79) – USA*

Primary energy consumption in Britain (1950-79)

Our second example in this chapter is one of the major industrialized nations in Western Europe – Britain. In one sense it is atypical in that it now has indigenous oil supplies off its shores. However, these supplies had not been exploited to any great extent in 1974, so that Britain was in no better position to cushion the effects of the Middle East oil crisis than any other Western European country. Since the marketing of Britain's oil is effectively controlled by world oil pricing agreements, this new source of energy has not really influenced the primary energy consumption pattern to any great degree. Its main effect has been an improvement in Britain's financial position through the oil revenues that have accrued to her government and the effective elimination of the cost of oil imports from her balance of payments.

The primary energy consumption figures for Britain are given in Figure 5.12 while Figure 5.13 shows the total energy consumption in graphical form.

In the decade to 1960, the average annual rise in energy consumption was 1.5 per cent, and, between 1960 and 1970, was roughly two per cent per annum. As in the USA, primary energy consumption actually fell between 1973 and 1975, but rose once again (at an average rate of 2.3 per cent per annum) over the period 1975-79. At the time of writing, the indications are that primary energy consumption in 1980 will be about eight per cent lower than the 1979 figure. This drop illustrates quite dramatically the effects of the current economic recession.

Unlike the USA, Britain has only recently begun to exploit her oil reserves. Also, the role played by coal in the total energy pattern has always been greater in Britain. Since the 1950s, however, oil and natural gas have replaced coal as Britain's main source of energy, and nuclear power has now also become an important contributor. This changing pattern is seen clearly in Figure 5.14.

Year	Coal	Oil	Natural gas	Hydropower	Nuclear power	Total
1950	204.3	20.6	–	0.9	–	225.8
1960	198.6	60.2	0.1	1.7	0.5	261.1
1970	156.9	132.3	17.0	2.3	10.1	318.6
1971	139.3	133.5	27.5	1.8	10.7	312.8
1972	122.4	143.1	39.0	1.8	9.8	316.1
1973	133.0	144.9	42.2	2.0	8.9	331.0
1974	117.9	134.6	50.4	2.1	10.7	315.7
1975	120.0	120.5	52.8	2.0	9.5	304.8
1976	122.0	118.4	56.1	1.9	11.4	309.8
1977	122.7	120.5	59.6	2.0	12.6	317.4
1978	119.9	123.0	61.8	2.1	11.9	318.7
1979	129.6	122.7	67.7	2.2	12.2	334.4

Figure 5.12 *Primary energy consumption (1950-79) – Britain (in mtce)*

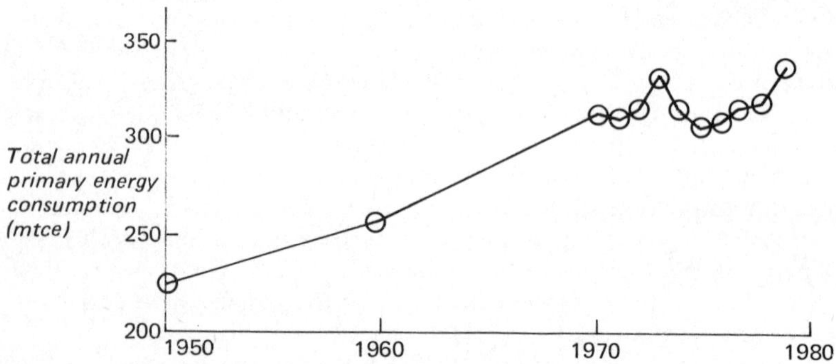

Figure 5.13 *Primary energy consumption (1950-79) – Britain*

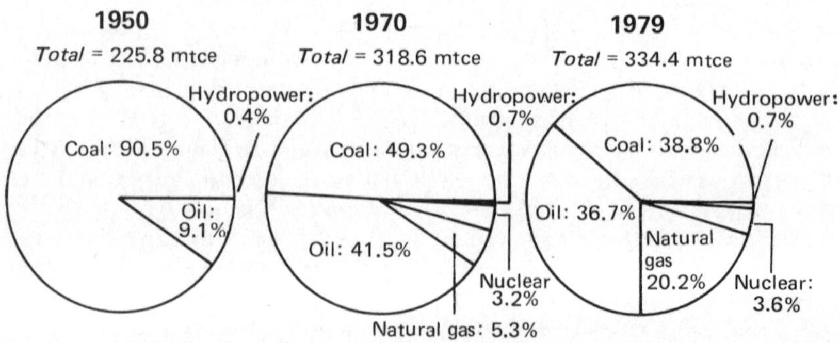

Figure 5.14 *Primary energy consumption by source (Britain) – the changing pattern*

Energy usage by sector

Figure 5.15 compares the energy usage in Britain in the different consumer sectors in 1972 and 1979.

	1972	1979
Residential (domestic)	24.8%	26.7%
Commercial and others	12.6%	12.9%
Industry	41.8%	37.6%
Transport	20.8%	22.8%

Figure 5.15 *Primary energy usage (Britain) by sectors*

By comparing Figure 5.15 with Figure 5.8, it is apparent that the residential (or domestic) market in Britain takes a much bigger share of the energy 'cake' than in the USA, while transport takes a smaller share. The residential sector in fact accounts for almost half the natural gas consumed, roughly 40 per cent of the solid fuel, and a similar percentage of all electricity generated; only about five per cent of oil is consumed in the domestic sector, however. As in the USA, the transport sector is almost totally dependent on oil.

Generation and use of electricity

From the 1950s through to the early 1970s, electricity consumption in Britain increased at an average rate of 6.8 per cent per annum. Consumption actually fell back between 1973 and 1975, but has since risen again. Between 1975 and 1979, the average rate of increase was 2.6 per cent per annum, but preliminary figures for 1980 indicate that consumption had fallen by about five per cent from its 1979 level. This decrease can be almost completely attributed to the recession in the industrial sector; electricity consumption in the iron and steel industry, for instance, dropped by almost 25 per cent from the 1979 figure!

Britain's installed generating capacity virtually doubled every ten years up to the early 1970s, when it began to level off after the Middle East oil crisis because of the resulting fall in demand for electricity. This levelling off was achieved by premature closure of old, relatively inefficient stations as the new stations ordered in the 1960s came into service. In 1979, Britain had a total generating capacity of 67 GW – more than enough to cope with her peak demand of roughly 50 GW. Like most other industrialized countries during the late 1970s, Britain ended up with several new stations that had been planned before 1974 being brought into service at a time when they were not really needed because of the fact that demand was much lower than had been anticipated.

In 1979, it is estimated that over 36 per cent of Britain's total primary energy input was used in the generation of electricity. This amounted to roughly 122 mtce, and a breakdown of this input by source is given in Figure 5.16.

Primary fuel	Amount (mtce)	Per cent of total
Coal	88.8	73.0%
Oil	16.1	13.2%
Natural gas	0.9	0.7%
Nuclear power	14.4	11.8%
Hydropower	1.6	1.3%

Figure 5.16 *Use of primary fuels in the generation of electricity (1979) – UK*

As can be seen, the pattern of primary fuel usage for electricity generation in Britain differs significantly from that in the USA. Almost three quarters of Britain's electricity is generated from coal, with less than one per cent being produced from natural gas. Hydropower also makes a very small contribution in comparison with the USA, although hydroelectric pumped storage is being used to an increasing degree in order to store electricity at times of low demand, thus enabling large, highly efficient power stations to generate continuously.

Energy consumption per capita

The energy consumption pattern follows the same trend as all the previous cases that have been quoted. Although the per capita figures are not as high as those for the USA, they are significantly higher than the overall world figures – as might be expected for a major industrialized nation. As can be seen from Figure 5.17, the Middle East oil crisis had only a temporary effect on the historical increase in consumption. The recession which has spanned the beginning of the 1980s seems likely to cause a further setback, although at the time of writing, comprehensive figures were unavailable for this period.

	1970	1973	1974	1975	1976	1977	1979
Population (millions)	55.42	55.91	55.92	55.90	55.89	55.85	55.88
Energy consumed per capita (tonnes of coal equivalent)	5.75	5.92	5.65	5.45	5.54	5.68	5.98

(Population figures from UK government statistics. 1979 figures are estimates.)

Figure 5.17 *Per capita primary energy consumption (1970-79) – UK*

CHAPTER SIX

Use of Energy – Possible Future Scenarios

The only safe conclusion that can be drawn from any technological forecast is that the further it looks ahead, the more likely it is that it will be wrong. In addition, it is probably true to say that the agency that commissions or carries out the forecast will predetermine (to a considerable extent) the eventual outcome, since it not only chooses the model on which the forecast is to be based, but also decides what the underlying assumptions will be, which figures will be fed in, and so on. An example that springs to mind is the way in which our electricity generating boards and power companies try to cater for likely future demand by attempting to predict what this demand will be. These have the thankless task of trying to make sure (a) that they have enough plant available to meet all possible requirements, and (b) that they do not build any plant that they will not actually need. Since our boards have to plan at least ten years ahead (the time it takes to order, build and commission a new station) they try to make their forecasts self-fulfilling – a policy that can come to grief when there is a sudden unexpected change in consumption pattern as in the period following the 1974 Middle East oil crisis (see Chapter 5).

Technological forecasts serve a number of extremely useful purposes, however. In many cases, a forecast is produced with the sole purpose of modelling the future behaviour of an existing economy, energy usage pattern or some other important social parameter in order to investigate 'in advance' whether or not present trends will be acceptable or even feasible in the years to come. Such forecasts therefore enable possible future problems to be identified in advance. An analysis of this type often provides indications of the kind of actions that have to be taken *now* in order to avert serious problems in the future. An alternative approach is to use a model which deliberately implies a change in attitude or way of life in order to identify the range of options available. Neither technique succeeds in predicting what the future will actually be; rather, between them, they identify the extreme possibilities. In this, they are valuable exercises, since they provide us with detailed analyses on the basis of which future policy may be determined.

In this chapter, we will begin by taking a general look at the principles that underlie energy policy forecasting, and will then examine two specific scenarios. The first, the so-called *historic growth* scenario (also known as the *unrestricted growth* or *business-as-usual* scenario) attempts to look ahead by extrapolating the trends that have been evident during the last 30 years or so into the future. The second, the so-called *zero growth* (or *low growth*) scenario, assumes that the present policy of unrestricted growth is wrong, and that energy usage patterns *must* change. By looking at the underlying assumptions and implications of each scenario, we will (we hope) be in a much better position to define the possible limits to future energy usage that will be needed for the discussion in Chapter 7.

The basic elements of any energy policy

The prime objective of any energy policy is to ensure that total energy demand is matched by total energy supply. Unfortunately, this objective can sometimes be difficult to achieve. In December 1978, for example, France experienced a general blackout as a result of a peak demand for electricity that could not be met by the available generating plant. This situation does not occur very often, but when it does, it is either a result of extreme demand circumstances or unforeseen operational difficulties or, in some cases, inadequate planning.

It is, however, not sufficient that total demand and total supply simply be balanced, since an adequate energy policy should also aim to provide the required energy 'mix'. This involves analysing the energy supply situation from both sides. On the one hand, existing and potential sources have to be examined with a view to deciding how future supplies can be guaranteed. This approach must then be reconciled with the opposite approach of trying to identify the future needs of the various energy-consuming sectors. Both approaches involve a certain degree of risk, since an attempt is being made to predict the future in a world in which very few nations are complete masters of their own destinies. Let us illustrate this by considering the events of the last 30 years.

During the 1950s, the availability of cheap oil (coupled with the application of Keynesian economic policies) paved the way for an unprecedented increase in standard of living in the countries of the Western world. As a result, the motorcar became commonplace, petroleum-based chemical and plastics industries flourished, and we became quite used to buying goods wrapped in at least two layers of cellophane. In many countries, traditional energy sources such as coal effectively went into decline, since oil was cheaper, more versatile and cleaner. There is no doubt that, *at the time*, this change in the energy usage pattern made good economic and social sense – after all, coal is not really suitable for running motorcars!

The 1974 Middle East oil crisis that upset this idyllic state of affairs was basically economic in its effect since it was not the *amount* of available oil that changed overnight but rather the *price* of that oil. The industrialized nations of the West were not able (or willing) to foresee that such a crisis was virtually inevitable sooner or later, and, as a result, all energy policies suddenly looked rather frail. To date, we seem to have coped with rising oil prices by being prepared, in the short term at least, to accept a lowering of our standard of living. The fact that our reserves of oil are (and always have been) finite has taken on a new significance, and nations are now having to plan ahead under a totally different set of conditions.

The criteria on which any energy policy is based are basically the same today as they have always been in the past. Apart from meeting the total demand with the correct mix of energy sources, a satisfactory policy should not involve too great a capital investment, should not result in large-scale unemployment and should not be based on obsolete technology (or on science fiction!). In addition, it should be *flexible*, so that national changes in energy usage and international changes in circumstances can be accommodated. In short, the overall policy should make economic and social sense, and should leave the door open for the emergence of new technologies and new energy options.

Unfortunately, the picture presented above is only too clearly seen to be an over-simplification. The ability of any energy policy to cope with change, either internally or externally initiated, is, in practice, very limited. For example, it is virtually impossible to bring about major technological changes in a short time. Nuclear power has taken over 20 years to reach the stage where it contributes roughly three per cent of the total energy budget. It may take another 20 years before nuclear power can make a major contribution, since to do that will probably require the development of commercial fast reactor systems (see Chapter 9). Another factor which contributes to this long lead-in time is that of social acceptance. This has certainly influenced the development of nuclear power, and is probably also responsible for the delay in other developments, such as the electric car. For similar reasons, it also takes time to replace obsolete existing technologies. The coal-fired power station at Battersea in London, for example, is still being kept in operation despite the fact that, in terms of electricity generation, it is obviously no longer economic. The reason for not closing it down is that it also supplies most of the houses in the area with a cheap source of hot water for domestic heating – something that the households involved would be very loath to forgo.

The dramatic rise in oil prices that has taken place since the mid-1970s has encouraged many governments to take a serious look at alternative energy resources – perhaps for the first time. (It can be argued, of course, that they are still not looking seriously enough.) Even though the basic technology required to exploit many of these resources is fairly well established, it is still likely to take several years (or even decades) before economically proved systems can be introduced. At the moment, the best that can be hoped for is that these options will be included in future energy forecasting, since, at the present time, contributions from such sources cannot be regarded as completely reliable (an essential feature of a main-line source of energy).

Let us now take a look at the two extreme scenarios mentioned above, namely, historic growth and zero growth.

The historic growth scenario

The basic assumption underlying this scenario is that historic trends, established over many years, are likely to continue into the future.

On a personal level, this implies that we will continue to increase our material prosperity by, for example, buying more motorcars and basic household appliances. In addition, it implies that we will have reasonably full employment coupled with reasonably low inflation and interest rates.

It is also assumed that major political and economic institutions will continue in their traditional roles. Problems such as inequality of income and opportunity will be effectively circumvented by real increases in personal wealth (the problem will not go away, but, with everyone feeling better off, the injustices will not appear to be so severe!).

A further assumption is that industry will continue to be a major generator of national wealth. For this to happen, it must be assumed that the rest of the world will also develop according to the same scenario, since any major change in the attitudes or economic policies of a nation's trading partners would seriously affect that nation's economic growth.

In short, the historic growth scenario conforms to the social expectations that have been built up over the past 25 years.

Almost all economic activity depends in some way on energy, and there appear to be definite links between economic growth, standard of living and the consumption of energy. Any rise or fall in economic activity tends to be associated with a similar change in consumption of energy. As we have already seen in the previous chapter, the effect of the change in oil prices in 1974 had a dramatic effect on the rate of growth of energy consumption, particularly in the industrialized countries of the non-communist world. The effect was also noticed in a change in the gross national product (GNP) of many of the countries most dependent on Middle East oil supplies.

There are a number of indicators of national wealth that are linked with energy consumption, and the GNP is just one. (It can be defined as the total money value of all final products such as consumer goods and services plus total gross investment.) Although it is sometimes considered dangerous to link GNP very closely with energy consumption, the changing patterns of the two certainly show a remarkable similarity. Figure 6.1 demonstrates this quite clearly for the USA.

At first sight, the historic growth scenario seems to ignore the fact that economic growth tends to move in cycles. However, the general upward trend that is apparent in Figure 6.1 certainly seems to justify the basic assumption that underlies the scenario, namely, that setbacks in economic growth (or rising energy consumption) are only temporary, and that the overriding historic trend will be maintained. There is considerable evidence to support this hypothesis.

Between 1920 and 1950, the average rate of increase in energy consumption in the non-communist world was roughly 1.8 per cent per annum, and, in the 20 years from 1950, this increased to almost five per

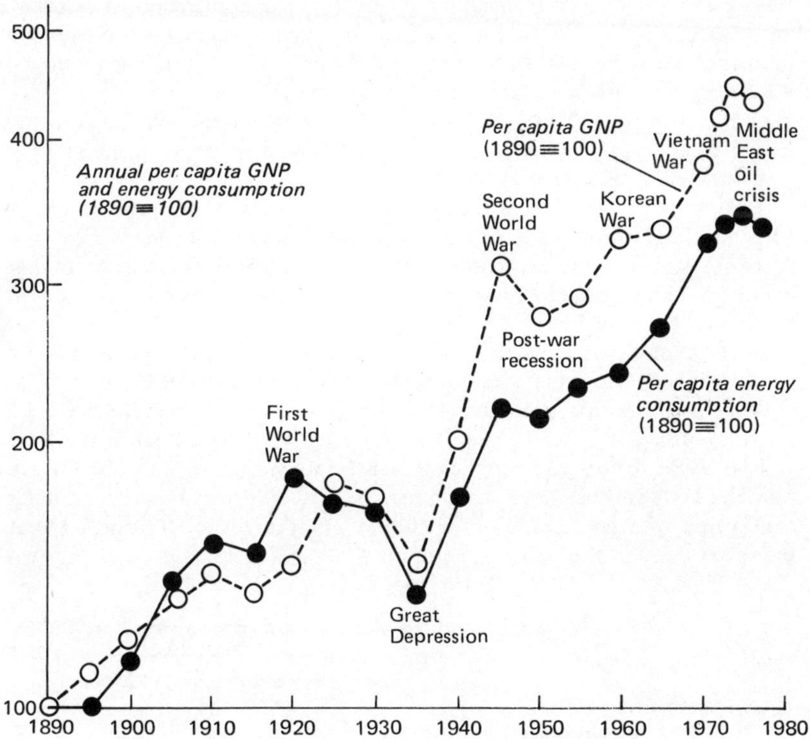

Figure 6.1 *The historic growth of per capita GNP and energy consumption in the USA*

cent per annum. Since 1975, the rate of increase has dropped, but is still around three per cent per annum. Taking the whole period from 1920-79, the average figure is roughly three per cent per annum. The world as a whole actually consumed over three times as much energy in 1979 as it did in 1950 – equivalent to an annual increase of almost five per cent. Even in Britain and the USA, energy consumption has still been rising at between two and three per cent per annum since 1975. Thus, it appears that the upward trend is virtually unstoppable, and this is the basis of the historic growth scenario. (It is, of course, true that the upward trend *might* be halted if people were prepared to make a drastic change in their attitudes and/or lifestyle, but the historic growth scenario makes the not-unreasonable assumption that this is highly unlikely. Recent events in Canada, where the federal government was thrown out of office after trying to introduce fairly modest energy conservation measures, certainly support this view.)

There are two other reasons for exploring the historic growth scenario.

Firstly, it is the one often assumed by governments and leaders of industry, and has in the past been the basis of much important government and industrial planning. Secondly, any serious analysis of future energy policy *must* examine the consequences of a continuation of historical growth. We cannot be sure that such growth will not be maintained, since, if the future is anything like the past, new unforeseen sources of energy may well appear on the scene.

Various estimates have been made of the rate at which energy consumption would rise under the conditions of the historic growth scenario. At the time of writing, the most commonly held view is that energy demand would increase at roughly the same rate as was experienced between 1975 and 1979 (2.5 per cent per annum for the USA and 2.3 per cent per annum for the UK). The current world recession, which is likely to influence 1980 energy consumption figures, is considered to be just another of those 'hiccups'. It is assumed that no deliberate efforts will be made to alter patterns of energy usage, and that vigorous national efforts will be made to increase supplies in order to keep up with demand. If we take the USA and the UK as examples, the historic growth scenario implies total primary energy consumption figures for the years 2000 and 2020 of the order given in the tables in Figure 6.2.

USA – historic growth rate of 2.5 per cent per annum

	1979	2000	2020
Total primary energy consumption (mtce)	2890	4854	7954

UK – historic growth rate of 2.3 per cent per annum

	1979	2000	2020
Total primary energy consumption (mtce)	334.4	539.0	849.5

Figure 6.2 *Likely future energy requirements in the USA and UK according to the historic growth scenario*

Similar tables can be drawn up for the growth in electricity demand over the same period. Again, it is assumed under this scenario that per capita use of electricity will increase steadily in the long term. The tables in Figure 6.3 are devised on the assumptions (a) that demand will grow at an average rate of three per cent per annum, and (b) that peak generating capacity will be roughly 28 per cent greater than peak demand (the overall safety margin with which most generating boards and power companies like to operate).

USA – historic growth in electricity demand

	1979	2000	2020
Installed electricity generating capacity (MW)	528,000	1,003,000	1,811,500
Total primary fuel input required (mtce)	900	1700	3090

UK – historic growth in electricity demand

	1979	2000	2020
Installed electricity generating capacity (MW)	67,000	118,060	213,235
Total primary fuel input required (mtce)	122	215	390

Figure 6.3 *Likely future electricity requirements in the USA and UK according to the historic growth scenario*

It is tempting to analyse the historic growth scenario in much greater detail at this stage of the proceedings, and we are sure that the reader has already formed an opinion as to whether or not it is feasible and represents a realistic view of possible future trends. However, it must be remembered that the scenario is being used purely in order to define the likely upper limits to growth. For this reason, we will refrain from further comment until the end of the chapter, when we will have had a chance to examine the 'lower limit' scenario.

The zero growth scenario

The concept of zero growth is understood by different commentators to mean different things. Until quite recently, 'zero growth' was taken to mean that the annual growth rate gradually fell during the forecasting period, until, at the end of the period under study, it was actually zero. In 1979, however, Gerald Leach and his co-workers (see Bibliography) published a fascinating study of possible future energy demands in the UK in which it was assumed that energy demand would remain roughly constant (and in one version of the scenario would actually *fall* slightly). For our present purposes, zero growth will be taken to mean that, at the end of the forecasting period, the total primary energy demand will be roughly the same as at the starting point. It is not anticipated that this assumption will invalidate the generality of the arguments and implications that follow, however.

The protagonists of the zero growth scenario argue that it is worthy of consideration for several reasons. Firstly, they maintain (with some justification) that the present trends in energy consumption in the industrialized nations of the world are taking us further and further away from a stable, sustainable way of life. Most of the world's natural resources are either limited or non-renewable (or both), so that continued growth seems certain to bring us up against natural limits at some future date unless completely new technologies can be developed in the meantime. In view of the inevitable time lag between the inception of an idea and the implementation of any resulting new technology, they argue that we should be *conserving* our valuable natural resources, *not* using them up at an ever-increasing rate.

Secondly, they argue that man will eventually come to realize that long-term survival is more important than increased affluence.

Thirdly, they maintain that we could make *much* more efficient use of energy than we do at present (although they admit that it might be more expensive to do so, at least in the short term).

In short, they believe that the whole pattern of modern Western life is wrong, and that the adoption of the zero growth scenario would at least point us in the right direction.

The supporters of the zero growth scenario believe that it would not necessarily preclude economic growth, might actually *create* more jobs, and, with its emphasis on greater efficiency, would greatly enhance the quality of life. With the resulting emphasis on the development of new technologies, and, in particular, the alternative sources of energy, the pressure on our fossil fuels would be greatly reduced (so they believe), and large-scale nuclear power programmes would be unnecessary.

As has already been demonstrated in the previous section, history is against this radical scenario. Attitudes and personal aspirations would have to change – or be changed – quite dramatically in order to give this scenario a ghost of a chance. To ensure its success, governments would probably have to introduce legislation which could be said to impose unrealistic restrictions on personal and economic freedom. For example, there might have to be a significant (compulsory) shift to public transport, with a consequent reduction in the use of private cars. Other measures that might have to be taken include severe tax disincentives against the use of electricity for space heating, comprehensive insulation of buildings, introduction of combined heat and power schemes, reduction in growth rates in the goods and public service sectors, and a general redirection of economic growth away from energy-intensive industries towards economic activities that require less energy – less emphasis on making things and more on offering marketable services.

The success of this scenario is perhaps even more critically dependent on the ability of any nation pursuing it to maintain its trading position with the rest of the world. It seems highly unlikely that an industrialized nation opting unilaterally for a zero growth scenario could possible survive for very long unless its competitors adopted similar policies.

Historic growth or zero growth?

Detailed descriptions of the two scenarios outlined above have been given for both the USA and the UK, and we have no intention of duplicating these excellent studies here. The interested reader is referred to the Bibliography. We feel that they are important because they demonstrate the likely consequences of two highly different sets of assumptions and policies, and because they present us with two extreme cases. Let us therefore try to decide which of the two scenarios is more likely to occur in practice.

If it were generally agreed that one or other of the scenarios was the more realistic and desirable, and made most economic sense, then that would obviously be the one most likely to succeed. Choosing the historic growth scenario, for instance, would require a concerted effort with a view to providing the large additional amounts of energy that would be needed. *Provided this could be done*, we would find that electricity would begin to play an even more important role in the total picture, and would probably account for the bulk of the growth of energy consumption in the domestic sector. A smaller proportion of the population would live in city centres, choosing instead to move into the suburbs. This population drift would create an increasing dependence on private transport. The pace of life, dictated by this scenario, would probably also result in a considerable expansion of air travel, both passenger and freight. Industry and commerce would continue to be prosperous and near full employment would be the norm. Industry's share of the primary energy budget would increase, and per capita energy consumption would continue to rise.

At the time of writing, it is our belief that nations like the USA and Britain could *not* sustain such unrestricted growth for any length of time without having to rely more and more on imports of already expensive primary energy resources. The other industrialized nations of the Western world seem likely to be in a similar position, so these resources will undoubtedly become increasingly more expensive, particularly if they continue to be depleted at a rate which is greater than the rate of discovery of new reserves (witness the eightfold increase in oil prices over the period 1973-80). The philosophy of unrestricted growth in the industrialized countries of the world also has serious consequences for the developing world. Given that the total energy 'cake' is limited, it seems inevitable that the disparity between the developed and undeveloped worlds would increase. (Roughly 2000 million people live in developing countries outside the Middle East, and consume, on average, just under one tonne of coal equivalent per year. This should be compared with approximately 13 tonnes of coal equivalent in the USA and almost six tonnes of coal equivalent in the UK.) Until comparatively recently, the provision of energy in a conveniently usable form was considered to be simply a matter of tapping the appropriate natural reserves. Now, however, it is accepted that energy resources are strictly limited – at least with present technologies. These limited reserves will, therefore, have to be shared out, perhaps more equitably than in the past. Thus the historic growth scenario, which is based on the *laissez-faire* energy policies of the past, is surely out of date.

Supporters of the zero growth (or low energy) strategy argue quite strongly that this scenario can also be consistent with improved living standards (warmer houses, more household appliances, greater car ownership and air traffic, greater industrial output, etc) and, in this sense, it differs little from the historic growth scenario. Where the two scenarios *do* differ significantly is in the amount of energy that is needed to support them. The zero growth scenario relies for its success on massive conservation programmes and significant changes in the efficiencies of energy usage. On the face of it, it appears to be the much more attractive proposition. However, as has been said earlier, public attitudes would have to change significantly, and a large increase in governmental control would have to be tolerated. More importantly, the costs of the large-scale conservation measures that would be essential for the success of the policy are difficult to estimate. A successful zero growth policy would require the widespread use of combined heat and power, district heating, heat pumps and domestic use of coal. At present fuel prices, however, most of these technologies are found to be uneconomic (particularly in the UK), and so the cost of implementing the policy could be high.

Our purpose in presenting these two extreme scenarios has not simply been to show that they predict alternative lifestyles for the future, but rather to show that neither is really practicable. As technology advances, we feel that the pendulum will inevitably swing towards *some* form of low energy growth scenario because this is the one which is most likely to enable the industrialized nations of the world to continue to prosper while also allowing the developing countries to continue to develop. In the meantime, we will probably have to be content with some kind of compromise – with some form of *technifix* scenario, as it is often described.

Various forms of technifix scenario which attempt to achieve the advantages and exclude the disadvantages of the two extremes have been proposed in recent years. Under these scenarios, energy growth is assumed to lie somewhere between the upper and lower limits, and new technologies are encouraged in order to meet this energy growth. At the same time, programmes of conservation measures which can be pursued without creating economic chaos are proposed. Figure 6.4 is a schematic representation of the various future scenarios, and it should be noted that the technifix case is deliberately left as a band of uncertainty. In the next chapter, we will explore the energy options likely to be available in Britain and the USA, and will return to this figure in order to see whether this uncertainty can be reduced.

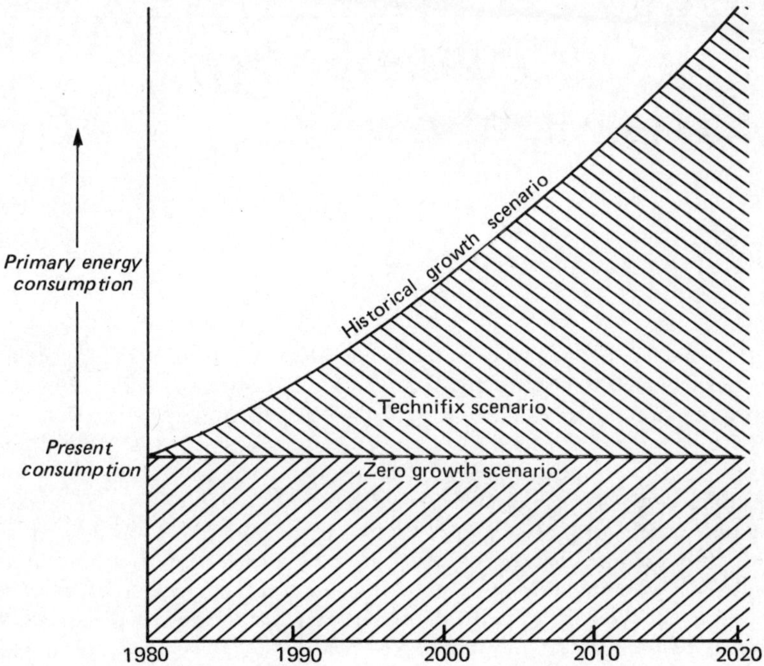

Figure 6.4 *Comparative sketch of different scenarios*

CHAPTER SEVEN

How Our Future Energy Needs Might Be Met

At the end of the last chapter, we deliberately refrained from attempting to predict specific figures for future energy requirements and confined the discussion to an examination of the two extreme scenarios, historic growth and zero growth. Since future energy consumption patterns will be determined by future availability of energy resources (as much as by anything else!) it seems sensible to delay any predictions until we have examined these resources in more detail.

In this present chapter, we will first take a look at our existing fossil fuels (coal, oil and natural gas), their reserves and their potential for future development. We will then do the same thing for the so-called *alternative* sources of energy (solar, wind, wave, tidal, hydropower and geothermal power) with a view to estimating the contribution that each might make towards the total energy budget. Finally, we will examine the USA and UK situations in order to build up an 'energy resource profile' for the year 2000; this will show that, even if they adopt a policy of low energy growth, both countries are likely to require significant amounts of nuclear power. Before we do so, however, let us try to explain what we mean when we ask the basic questions: 'How much of a resource is available?', and 'How long will it last?'

How much?

Official estimates of how much of a given resource exists differentiate between (a) *recoverable reserves* and (b) *total reserves*. Recoverable reserves are usually taken to include the amount of the resource that can be made available at an economic price using current technology. For example, the price of a barrel of oil in 1980 was around $35. If, instead, it had cost $350 per barrel, this would have been considered uneconomic. A maximum economic price for recovery is usually agreed upon at any given time, and all the oil that can be made available at that price is considered recoverable. Total reserves, on the other hand, is the term generally used to include recoverable reserves plus other known reserves which are uneconomic to exploit at the time. Both these estimates are important because, as fuel prices rise, the economic price of a given resource also tends to rise, so that a greater proportion of the total reserves can be considered recoverable.

How long?

The answer to this question ought to be relatively straightforward, but in practice this is far from the case. If a given resource were being used up at a steady rate, and the recoverable reserves were fixed, the answer would simply be the ratio of the recoverable reserves to the usage rate. However, as we have just seen, the economically recoverable reserves of a resource change as the economic climate changes. Furthermore, the rate at which each resource is used up also changes. Thus, the best that can generally be done is to quote the estimated lifetime of each resource assuming that the current usage rate will continue.

Fossil fuel energy resources

Coal

It is a generally accepted fact that coal was produced by the decomposition of primeval forests that flourished hundreds of millions of years ago. It has thus taken a *very* long time for our current reserves of coal to be created, and they must therefore be considered to be both finite and non-renewable. The world's coal resources are, however, vast, although it is estimated that over 90 per cent of them are located in only three countries – the USSR, USA and China. The total reserves of all types of coal (anthracite, bituminous coals, lignite, etc) have been estimated at between 8000 and 11,000* billion tonnes, although the *measured* reserves are actually only about 1300 billion tonnes, of which only 700 billion tonnes are currently estimated to be economically recoverable.

In 1979, world annual consumption of coal stood at just under 3000 million tonnes. If this consumption rate were to remain steady, then the presently designated recoverable reserves would last for just over 200 years. Although world consumption of coal almost doubled in the period 1950-1979, coal's share of the total primary energy demand actually *decreased* from 57 per cent to 28 per cent over the same period as coal was systematically replaced by oil and natural gas. Oil is easier to transport than coal, occupies less space and yields more energy per tonne. For a considerable time, it was also *cheaper*, since coal was only competitive with oil when it was consumed close to its source. This is reflected in the fact that the world's international trade in coal in 1973 was only 170 million tonnes. Most of this coal was imported by Japan and Western Europe, and nearly all of it came from the USA, Canada and Australia.

Since 1974, coal has recovered some of its lost popularity, and, as oil prices continue to rise, it is likely that international trade in coal will also increase. There is already evidence that coal is beginning to replace oil and natural gas in some industrialized countries (see Figure 5.3 for instance), and, as oil becomes steadily more expensive, it is expected that this trend

* Throughout this book, the word 'billion' is used in the American sense, ie one thousand million (1,000,000,000).

will continue unless vast reserves of as-yet-undiscovered oil materialize in areas of the world where OPEC can exert little influence.

In the next few decades, coal is likely to play an increasingly important role in the world's primary energy budget. Generating electricity from oil, for example, is already so uneconomical that some oil-fired power stations only recently brought into operation in the UK are already being 'mothballed', their role being reduced to that of emergency standby stations. Coal-fired power stations are therefore going to be a familiar landmark for many, many years to come.

Although coal will certainly continue to be burned (as coal) in the future, it will almost certainly also have to be used as a raw material for the manufacture of synthetic oil and gas. To produce synthetic oil, finely divided coal is passed through a series of reaction vessels in which it is gradually transformed into a mixture of vapourized hydrocarbons. These are then cooled and separated into two streams, one of which consists of 'synthetic gas' and the other of fractions which can be converted into 'synthetic crude oil' by a process known as *hydrogenation* (adding hydrogen to the molecules of the various substances present). To obtain synthetic gas as a primary product, the coal can be burned in an atmosphere of steam and oxygen. If this is done at a comparatively low temperature, a gas known as *town gas* is produced; at higher temperatures, a gas with a higher methane content having similar properties to natural gas is produced.

Coal in the USA

It has been estimated that the USA's recoverable reserves of coal stand at between 180 and 270 billion tonnes. In 1979, the USA's home consumption of coal was just over 600 million tonnes. At this rate of consumption, her recoverable reserves would last for roughly 300-450 years. In America, it is the electricity generation and industrial sectors that are most likely to see a significant shift towards coal and away from oil and natural gas in the years to the end of the century. In 1979, over 46 per cent of all her electricity was generated in coal-fired power stations, and the National Energy Act of 1978 seems certain to cause this figure to rise steadily as the use of natural gas for electricity generation is gradually phased out during the 1980s. Indeed, some commentators believe that up to 65 per cent of America's electricity will be generated from coal by 1995. On the basis of the 1979 figures, this would represent a 40 per cent increase in coal consumption – equivalent to an average two per cent per annum increase over the intervening period.

It is very difficult to estimate what the total level of coal production will be in the USA in the year 2000. There is no American equivalent to the British National Coal Board, which forecasts total production levels. Most American reports cover a range of possible future scenarios, and the emphasis is always on how much of each resource will be needed, rather than on what kind of level of increase is possible. However, the consumption of coal in the USA has increased by between three and four per cent every year since the Middle East crisis. In view of the increasing importance of coal in her economy, we believe that this rate of increase could be sustained through to the year 2000. An annual increase in production of 3.5 per cent would give a production rate of 1500 million tonnes in that year (ie the 1979 figure would be approximately doubled). If the USA continues to export up to ten per cent of its coal production, this would leave about 1350 million tonnes for domestic consumption. The US coal industry faces a number of problems, however, including the need for more miners, the need to introduce better production facilities, the need for increased investment and the need for more efficient pollution control. Indeed, the environmental impact of a major growth in coal production could be one of the most important factors, since roughly one half of all coal produced in the USA is at present obtained by open-cast mining.

USA — Likely annual primary energy contribution from coal by the year 2000: ~1350 million tonnes

Coal in the UK

Recoverable coal reserves in the UK are conservatively estimated at five billion tonnes and total reserves at 100 billion tonnes. Production of coal stood at roughly 130 million tonnes in 1979, and, at this rate of usage, the recoverable reserves would last for only 40 years! This perhaps surprisingly short lifetime will no doubt be increased considerably as the price of coal becomes more competitive and oil and natural gas become scarce commodities. The total known reserves that are generally believed to be *ultimately* recoverable stand at around 45 billion tonnes, or roughly 350 years' supply at 1979 extraction rates. This significant difference indicates just one of the problems that have to be overcome in attempting to evaluate recoverable reserves. In terms of the definition given at the beginning of this chapter, the UK only has 40 years of coal left, but, as is often the case, some of the total reserves become reclassified as 'recoverable' when the economic conditions change; it is basically a question of what people are prepared to pay.

In recent years, the British National Coal Board has produced two estimates for the future pattern of coal supplies. In 1980, an earlier figure of 170 million tonnes per year by the year 2000 was adjusted downwards to 155 million tonnes. An investment programme of some £10 billion is planned for new and modernized production facilities in order to meet this target. However, the plans also involve the phasing out of much of the existing capacity because of seam exhaustion, and these plans will undoubtedly be very closely examined by the workforce and miners' unions. If attempts to improve productivity and contain costs are thwarted, then even this moderate increase in production (an average increase of barely one per cent per annum) could be difficult to achieve.

UK — Likely annual primary energy contribution from coal by the year 2000: ~155 million tonnes

Oil

Oil is created from the decomposing remains of micro-organisms, animals and vegetation. The deposits that we find and use today were all accumulated over 100 million years ago, and, like coal, they must be regarded as both finite and non-renewable. It has been recognized for some years now (since long before the Middle East crisis) that the world's oil resources are being depleted at a rate that cannot be sustained. No very large oilfields (ie those with recoverable reserves in excess of 5000 million barrels) have been discovered since 1968, and to maintain the present level of production and reserves requires the discovery *each year* of two 'Alaskas', or as much oil as has been found in the UK sector of the North Sea in the last ten years.

If the world economy is to continue to grow in the long term, oil consumption will have to be reduced and other primary energy resources will gradually have to replace oil. As the BP booklet 'Oil Crisis . . . again?' puts it:

> The problem about oil is not that sometime in the next century the world is going to 'run out' of oil – as the media often put it. Rather, the problem is that within a decade, the world is going to have to switch from a supply which *grew* at 5½ per cent per year over the 50 years to 1973, to a supply which *declines*! (January 1980).

The picture can be made clearer if we look at the ratio of remaining reserves to production rate over the past 50 years or so. These are summarized in Figure 7.1 for the non-communist world (NCW) and the USA, and are based on 1979 estimates of reserves backdated to the year of discovery. The figures show that the reserves-to-production ratio (ie the likely remaining life of the reserves at current usage rate) has fallen steadily since 1940 in each case.

	Reserves-to-production ratio	
Year	**NCW**	**USA**
1930	99	40
1940	114	45
1950	95	35
1960	72	25
1970	42	15
1975	37	12

Source: BP

Figure 7.1 *Oil reserves-to-production ratio*

In 1979, the world's proven recoverable reserves of oil stood at ~90 billion tonnes, of which 56 per cent were in the Middle East. OPEC members are responsible for over 80 per cent of the non-communist world's reserves, and any oil scenario is obviously going to be influenced by their future pricing and supply policies; we have already seen the effects of their control over the price and availability of oil in the years since 1973.

Oil in the USA

In 1979, the USA consumed nearly 28 per cent of the world's oil production while, at the same time, producing only 15 per cent herself. The USA is therefore one of the largest net importers of oil in the world (46 per cent of her annual consumption currently being met by imports) and will presumably remain so. Since indigenous US reserves are likely to last for such a short time (see Figure 7.1), it is possible that her imports of oil will increase during the 1980s and 1990s even though her total demand for oil will probably fall. In the long term, world production is expected to decline, and, despite the bargaining power of the USA, it is unlikely that her primary energy input from oil will be greater in the year 2000 than in 1979. The USA has vast deposits of oil in the form of oil shales, the largest being located in Wyoming, Colorado and Utah. At present, producing oil from oil shales is considered a high-risk venture; several of the proposed extraction methods require enormous amounts of cooling water, and this is seen as a distinct disadvantage because of the semi-arid nature of the regions in which most of the shale deposits occur. Techniques which avoid the large-scale use of cooling water are currently being developed, and only time will tell whether they will eventually become economic. As with coal, the exploitation of oil shales will require careful environmental control in what is still a relatively unspoilt area, and oil from this source is unlikely to make a significant contribution before the year 2000.

USA — Likely annual primary energy contribution from oil by the year 2000: ~1350 mtce

Oil in the UK

In 1980, the UK became self-sufficient in oil thanks to supplies from the British sector of the North Sea, where production platforms of the type shown in Figure 7.2 are now a familiar sight.

This self-sufficiency is a luxury which will probably be short-lived, however, since most forecasts suggest that production will decline after the mid-1980s. At present rates of consumption (~ 120 mtce per year), the recoverable reserves in the UK sector of the North Sea would last for between 30 and 50 years (estimated recoverable reserves being somewhere in the range of 4000-7000 mtce). Although demand for oil in the industrial and domestic sectors is likely to fall, non-energy uses will keep the overall demand level growing unless this is halted by increasing prices. Since the UK is not in the same strong bargaining position as the USA, it will be difficult to meet any shortfall in supplies with imports, and the British energy usage pattern might have to change significantly in order to cope

Figure 7.2 *A North Sea oil production platform*

with this. We expect that, by the year 2000, the UK will not be able to call on more than 100 mtce of oil per annum.

UK — Likely annual primary energy contribution from oil by the year 2000: ~100 mtce

Natural gas

This is the more general term given to a mixture of (mainly) hydrocarbon gases found in porous reservoirs in sub-surface rocks. Its origins are uncertain, but are thought to be similar to those of oil. The predominant gas is methane, which usually constitutes over 85 per cent of the total. *Associated gas* is usually found in conjunction with oil, and *non-associated gas* is apparently unrelated to oil. In the northern North Sea, for example, oil and natural gas are often found together, whereas in the southern North Sea, only gas is found.

In 1979, the estimated recoverable world reserves of natural gas stood at roughly 100 thousand mtce and world natural gas consumption in that year was roughly two thousand mtce. This implies a lifetime of about 50 years at this consumption rate. (Incidentally, a similar calculation for the non-communist world alone produces much the same result.) Natural gas is in a similar position to oil in that it is a highly versatile energy resource with a strictly limited life. In the short term, natural gas will probably be used to replace expensive oil wherever possible, and world consumption is therefore expected to continue to rise. By the turn of the century, however, production will probably have passed its peak and will be on the decline.

Natural gas in the USA

The published recoverable reserves of natural gas in the USA were seven thousand mtce in 1979, and consumption in that year was roughly 700 mtce. This implies that the USA has only about ten years' supply left if this rate of consumption is maintained. Commentators are already talking of an increasing shortage of natural gas, particularly in non-producing areas of the USA. Unless large new reserves are found, it is obvious that the present consumption rate cannot be maintained to anything like the year 2000. Even the most optimistic forecasts suggest that, by the turn of the century, production of natural gas will be down to about 60 per cent of its present levels, and this allows for substantial new reserves being found in the meantime. At the other end of the scale, some commentators believe that deposits will have been so depleted by the year 2000 that natural gas will only be available for purposes for which no other economically viable raw material is available. On present evidence, it would be irresponsible to predict a production rate for the year 2000 higher than about 300 mtce (even allowing for new finds).

USA – Likely annual primary energy contribution from natural gas by the year 2000: ~300 mtce

Natural gas in the UK

In at least one way, the UK is more fortunate than the USA with regard to its natural gas reserves – they were only discovered comparatively recently, and are still being discovered in the North Sea. Natural gas in the UK has a history of less than 30 years whereas, in the USA, it has been exploited for over 100 years.

In 1979, 68 mtce of natural gas were consumed, all of which came from the UK sector of the North Sea. Recoverable reserves are estimated to be between 900 and 2100 mtce, and, at the present rate of consumption, these would last for between 13 and 30 years. Most energy forecasts are based on the assumption that the UK will be able to continue consuming natural gas at roughly the present rate at least into the 1990s, although there is no guarantee that supplies will last much longer than this. Thus, by the year 2000, natural gas supplies could well be declining sharply, and we would expect a production rate of only 40 mtce by then.

UK – Likely annual primary energy contribution from natural gas by the year 2000: ~40 mtce

Alternative sources of energy

It is apparent from the above discussion that, although fossil fuels will continue to play a dominant role in meeting our future primary energy requirements up to the turn of the century, they are rapidly being depleted and will eventually have to be replaced by alternative energy resources. Coal is the only non-renewable energy resource with a projected lifetime long enough for it to be used in increasing amounts by the industrialized countries of the world well into the 21st century. The USA, with its reserves of oil shales, and Canada with its oil sands, may well be in a privileged position *after* the turn of the century, but, in the short term, these resources will probably only have a marginal effect on the total primary energy supply for those nations.

Solar energy

Solar energy has provided, directly or indirectly, almost all the sources of energy that have been produced on earth since its formation 4600 million years ago. Through the slow action of heat, pressure and ageing, plant and animal substances have been turned into coal, oil and natural gas. The amount of energy reaching the earth's surface from the sun is vast; on a global scale, the solar energy that arrives in just *two weeks* is equivalent to the total energy stored in all the earth's known reserves of coal, oil and natural gas. Nearer home, the solar energy that falls on an area smaller than that of the American State of Massachusetts could, if converted at only 20 per cent efficiency, meet the electricity requirements of the whole of the USA. Even in the UK, with its relatively high population density, the sun supplies many times more energy than is actually consumed.

There are, however, two disadvantages of solar energy. Firstly, the sun's energy is *diffuse*, ie it is spread out rather thinly. Secondly, it is *intermittent*, since the sun shines only during the day, and, even then, is often obscured by clouds, so that its energy must be stored until needed. Despite these intrinsic disadvantages, research into methods of harnessing solar energy is advancing along two quite different fronts – *using solar energy to provide heat* and *converting solar energy to electricity*.

Solar energy as heat

The main research efforts are in the areas of solar water heating, space heating and, to a lesser extent, biofuels. Figure 7.3 shows the various components of a solar water heating system. Here, the sun's rays are trapped by the collector plate and used to heat water which is circulating through the system. The heated water is then piped to storage tanks to provide domestic hot water. At present, solar water heating systems are hardly cost-effective, except perhaps against electricity, but their use throughout the world is growing. In the USA, almost one quarter of the total energy consumption is related to space heating, water heating and air conditioning. It is not difficult to envisage a modification of the scheme shown in Figure 7.3 whereby the hot water could be circulated through radiators. With regard to air conditioning, solar energy can be harnessed for cooling purposes by using solar heat to replace the gas flame in conventional air conditioners. It is expected that these relatively new

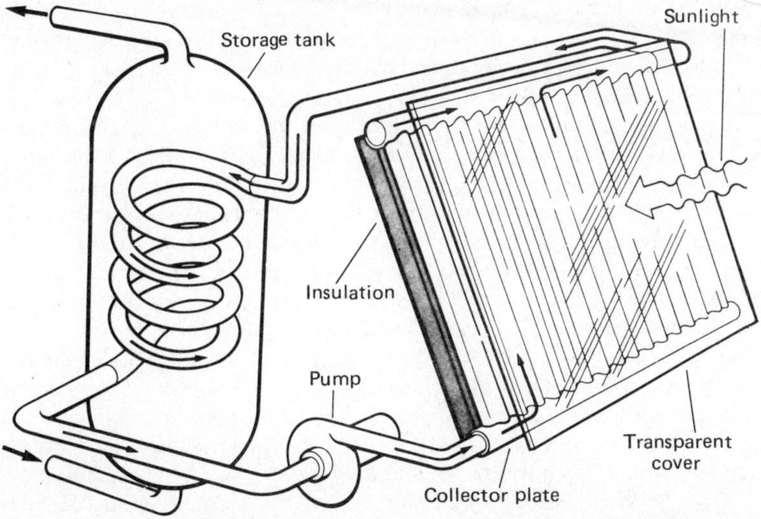

Figure 7.3 *A solar water heating system*

technologies will be developed over the longer term, but will not make a significant contribution before the beginning of the next century.

Photosynthesis is yet another way of converting solar energy to a useful form. This has been used by nature for millions of years to produce the energy required for plant growth. This *biomass*, as it is called, could be processed to yield gaseous and liquified fuels, and the commercial exploitation of biofuels could, in the long term, become economically attractive. For instance, it has been estimated that from each one per cent of the UK land area devoted to an 'energy crop', approximately one mtce per annum of biofuels could be obtained. (Incidentally, the crop could also possess a usable food component.)

Solar energy as electricity

There are a number of ways, both direct and indirect, by which solar energy can be converted into electricity. One direct method which has been used extensively in the NASA space programmes is the use of *solar cells*. In certain substances, the absorption of light creates an electrical voltage that can be used to generate electricity in an external circuit. This process, known as the *photovoltaic effect*, is also used in some kinds of photographic light meters.

A large array containing several million solar cells could be used to generate low-voltage direct current, which, after conversion into high-voltage alternating current, could either be fed directly into a national grid system or stored for future use. At present, the cost of generating electricity in this way is prohibitively expensive (roughly 50 times more expensive than conventional methods), but a great deal of research into developing new and cheaper types of solar cells is currently being carried out, and a technological 'breakthrough' may well be on the way – possibly within the next ten or 20 years.

A possible alternative to land-based solar power stations, which would occupy relatively large areas, is that of *solar power satellites*. Here, arrays of solar cells would be mounted on a satellite system which would be put into a synchronous orbit around the earth (22,300 miles above the surface). The satellite could generate electricity almost continuously, and it has been proposed that this be fed to microwave generators and beamed back to earth to a receiving antenna which would reconvert the microwave energy into electricity. Such schemes would benefit from the current advances being made in solar cell technology, and could be a viable source of base-load electricity in the 21st century. The important factors bearing upon the success of such schemes appear to be political, economic and environmental rather than purely technical. The development costs, for example, would be enormous, and could probably not be contemplated by a relatively small country like the UK on its own. A joint European venture, or, perhaps, a partnership between the USA and other nations, seems to represent the only way in which the smaller industrialized nations of the world could benefit from this 'satellite technology' in the foreseeable future.

A second possible technique for converting solar energy into electricity, known as *ocean thermal energy conversion* (OTEC), is shown schematically in Figure 7.4. This makes use of the fact that the ocean water temperature in the tropics falls from roughly 25°C near the surface to roughly 8°C at

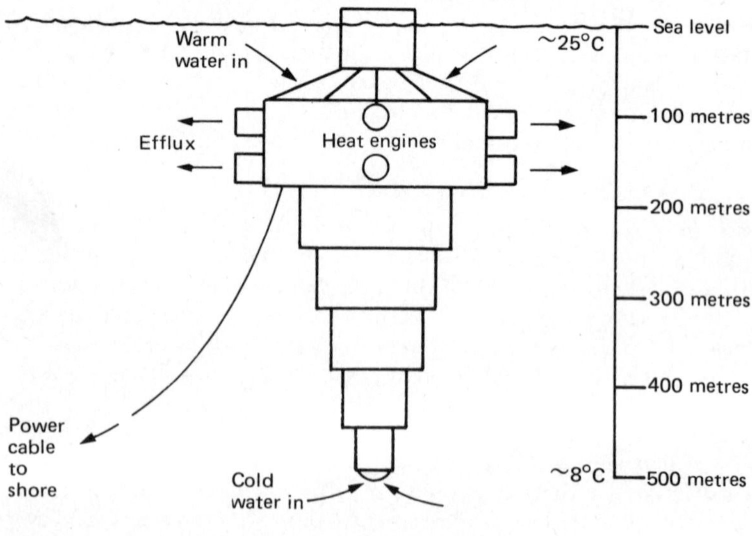

Figure 7.4 *The OTEC system*

a depth of 500 metres. It is therefore possible to use hot water taken in at the surface and cold water taken in at such a depth to drive heat engines and generate electricity. Because of the extremely small temperature difference, the efficiency of the heat engines would be very low, however (only a few per cent), so vast quantities of water would have to be passed through the system in order to produce electricity on a commercial scale.

The potential energy sources based on solar energy discussed so far are all possible energy sources for the future. However, with the exception of water heating (and perhaps some contribution to space heating), it would be reckless to plan a short-term future energy budget which relied on technologies which are not yet economic, and which (in some cases) have associated political and environmental problems that have still to be resolved. Our estimate of the potential contribution of solar energy to the overall energy supply pattern in the year 2000 is based on the assumption that it might, just conceivably, be equivalent to providing all domestic hot water from solar energy by that year.

USA – Possible annual primary energy contribution from solar energy by the year 2000: ~60 mtce

UK – Possible annual primary energy contribution from solar energy by the year 2000: ~10 mtce

The wind and the waves

Thermal gradients in the atmosphere produced by the incidence of solar radiation on the earth's surface cause the movement of large masses of air, and it is this energy of motion that windmills attempt to capture. The wind, in turn, produces waves on the surface of the sea, and, where the fetch of water is large (~ 100 miles or more), the potential power in these waves can be several thousand watts for each square metre of surface area.

Energy from the wind

The basic principle behind wind power is quite straightforward, and has been used for centuries to convert the energy of the wind into usable mechanical energy (the rotation of windmill blades). This mechanical energy can also be used to drive generators and thus produce electricity. Figure 7.5 shows a range of different windmill designs which have been proposed for use in electricity generation.

Compared with conventional energy conversion systems, aerogenerators must be classed as 'dilute' sources, since the equipment needed to produce useful amounts of power needs to be large, requires large amounts of constructional material, and may therefore be expensive. For example a 3.7 MW (peak generating capacity) prototype aerogenerator with 60-metre diameter rotors is currently being developed in the UK; *one thousand* of these machines would be needed to replace *one* modern 2000-MW fuel-burning power station.

Aerogenerators have to be custom-designed for a given wind climate if they are to be used efficiently. The windiest sites have to be identified,

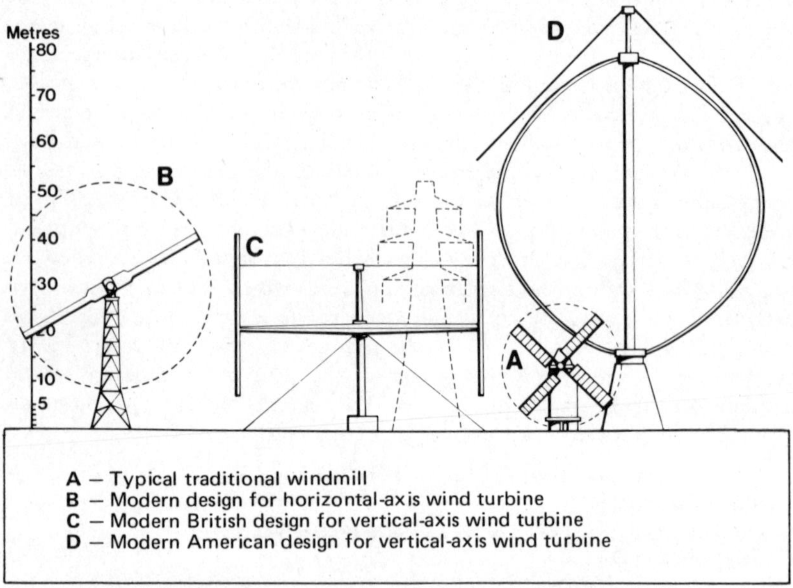

Figure 7.5 *Schematic representation of different windmill designs*

and, for the British Isles (for instance), this inevitably means those sites around the coast. In general, the best sites do not coincide with areas of maximum industrial or domestic demand for electricity. More often than not, they tend to lie in remote areas of great natural beauty. A one-megawatt aerogenerator would have to be about the same height as a 400-kV electricity pylon (see Figure 7.5), so that a large concentration of such aerogenerators could give rise to considerable environmental objections.

An alternative system would be to place the windmills out at sea. This has the immediate advantage that wind speeds over the sea are higher than those over the land. The capital costs of such schemes would be significantly greater than for an equivalent land-based scheme, however, because massive support structures would have to be included in order to cope with the hostile marine environment; at the time of writing, it is not clear whether the higher energy uptake at sea would offset the higher overall cost.

Although electricity from wind energy would be intermittent, its overall availability would correspond quite well with demand, since wind power levels are (in general) higher in winter than in summer, and higher during the day than at night.

We feel that the greatest potential for wind-generated electricity lies in remote areas, well away from the national grid systems, where it would be very expensive to provide centrally generated electricity. However, as other energy sources become relatively scarce and more expensive, there could well be a significant return to this ancient source of power, despite the relatively low load factor that could be achieved. Using suitable storage systems such as hydroelectric pumped storage (see below), wind power could provide relatively cheap top-up capacity for use in conjunction with more conventional base-load systems.

Wind power in the USA

In the USA, several wind power research programmes are now well under way. These are studying such things as wind energy conversion technology, wind characteristics and wind energy system designs, but it will probably take at least ten years for the true potential of wind power to be accurately assessed. It is, therefore, difficult to see wind power making more than a marginal contribution to America's total electricity requirements before the year 2000, but, during the first few decades of the 21st century, it could well become an important energy source if the various systems currently being studied turn out to be technically feasible and economically viable.

USA – Possible annual primary energy contribution from wind power by the year 2000: ~20 mtce

Wind power in the UK

At the beginning of 1981, the Central Electricity Generating Board (CEGB) in Britain announced that they were looking for a site on which to construct a prototype wind generator so that the future potential of aerogenerators could be assessed. By the end of the 1990s, they believe that they will know whether wind-generated electricity has a significant role to play in the UK.

Several estimates have been made of the potential contribution of wind power to the UK primary energy supply pattern for the year 2000. The high figures naturally come from the proponents of wind power, and those who believe that large storage systems (such as using 'empty' oil and gas strata for the storage of compressed air) can be developed, produce the highest estimates. These latter figures range from 30 mtce to 90 mtce, but we feel that a figure of about 15 mtce is more realistic in view of the environmental problems involved and the uncertainties in the economics of large-scale storage systems.

UK – Possible annual primary energy contribution from wind power by the year 2000: ~15 mtce

Energy from the waves

It has long been recognized that waves contain large amounts of energy, but it is only during the last decade that serious research into possible methods of extracting this energy, and using it to generate electricity, has been initiated. Wave power does not appear to feature in any USA energy scenarios, and so we assume that it will make no significant contribution to that nation's energy budget before the turn of the century.

Wave power in the UK

As waves move across the surface of the sea, they cause the water near the surface to undergo roughly circular motion. The energy associated with this motion can be tapped in a variety of ways, and Figure 7.6 shows an artist's impression of two possible wave energy devices that have been suggested by British designers.

In the 'Cockerell Raft' system, the movement of the waves causes the structure to flex, thus causing pistons to move to and fro in their cylinders. The resulting energy can be used to drive a generator.

In the 'Salter Duck' system, a row of specially shaped floating chambers impede the progress of the waves, and, on absorbing energy from the waves, begin to rock up and down. This mechanical energy can then be used to generate electricity by, for instance, pumping water through turbines.

It is expected that, by the mid-1980s, we will be in a position to know whether wave power is going to be a serious contender in Britain's energy stakes. The situation at the time of writing is that several different systems are under investigation, none of which has yet made the economic breakthrough that is necessary for it to achieve widespread use. The impression gaining ground among most impartial observers is that wind power is now looking rather more promising than wave power (the reverse of the

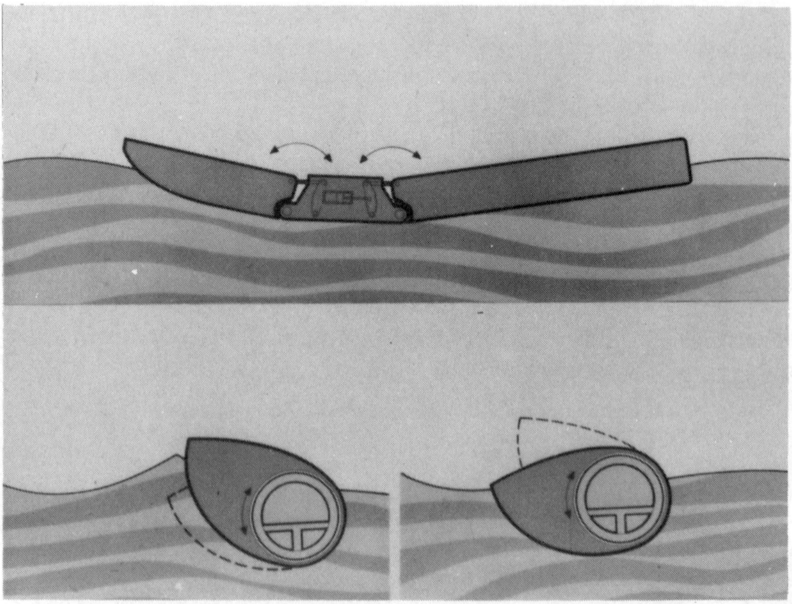

Figure 7.6 *Artist's impression of the 'Cockerell Raft' (top) and 'Salter Duck' wave energy systems*

situation a few years ago). A generous estimate of the possible contribution from wave power by the year 2000 would be in the range 20-30 mtce.

UK – Possible annual primary energy contribution from wave power by the year 2000: ~25 mtce

Tidal power

The twice-daily rise and fall in the level of the sea that is produced by the rotation of the earth through the gravitational fields of the sun and the moon represents a vast potential source of energy. Unfortunately, nearly all of this energy will always be inaccessible to man because of the great technical difficulties involved in harnessing tidal power on a large scale other than in a few places where special conditions prevail – mainly estuaries or inlets where the tidal range is exceptionally high.

If a barrage is built across the mouth of such an estuary or inlet, the rising tidal waters can be trapped, thus producing a store of potential energy that can be used to generate electricity once the level of the open water beyond the barrage has fallen. Tidal power is therefore essentially the same as hydroelectric power (which is discussed in the next section).

The only full-scale tidal plant built so far is that on the Rance estuary on the English Channel coast of France. This plant began operation in 1966 with an initial capacity of 240 MW, and has since been enlarged to 320 MW.

One of the world's most promising tidal power locations is the Bay of Fundy region of northeastern USA and southeastern Canada. This area has the world's highest tides, with ranges up to 15 metres, and a total power potential of up to 30,000 MW. In Britain, the most favourable site is the Severn Estuary, for which plans have been produced at regular intervals since the 1930s. It is estimated that such a scheme could save up to ten mtce per year. Neither of these potential resources has yet been earmarked for immediate development, and, since it is likely that construction could take up to 20 years, tidal power will almost certainly make no major contribution to the energy budget of either the USA or the UK before the year 2000.

Hydroelectric power

The technology of hydroelectric power generation is well established, and such stations are in widespread use throughout the world. The basic principle on which they operate is relatively simple. Natural rainfall is collected in a high-level reservoir as it runs off its catchment area. This high-level reservoir is usually a naturally occuring lake (or an artificially created lake) which has been dammed at or near its outlet in order to increase its water level and storage capacity. A power station is built below the reservoir (sometimes at or near the dam, sometimes a considerable distance away), and the water is fed to this through a tunnel or pipes; it is then used to drive turbines which, in turn, drive generators. The annual rainfall and volume of the reservoir determine the amount of electricity that can be produced each year, and the height of the reservoir above the turbines determines the rate at which the electricity can be generated.

A typical small hydroelectric scheme, at Pitlochry in Scotland, is shown in the photograph in Figure 7.7. This scheme is built across a famous salmon river, the Tay, and the structure in the lower left third of the photograph is in fact an artificial salmon ladder designed to allow salmon to reach the upper lochs, where they have spawned for centuries past.

Hydroelectric power in the USA

The world's first hydroelectric power station was built at the Niagara Falls in 1895, the electricity generated from this scheme being transmitted a distance of 26 miles to the city of Buffalo. In 1978, hydropower contributed almost 13 per cent of the USA's primary energy input for electricity generation, with a total installed capacity of roughly 75,000 MW. There are still several areas in the USA that are suitable for hydroelectric development, and we expect that hydroelectricity could increase by up to 50 per cent of its 1978 figure by the year 2000 if the necessary capital investment were made.

USA – Possible annual primary energy contribution by hydroelectric power by the year 2000: ~175 mtce

Figure 7.7 *The hydroelectric station at Pitlochry in Scotland*

Hydroelectric power in the UK

In the UK, the potential for further hydroelectric power development is almost entirely limited to its use as a storage system for surplus electricity produced by continuously operated base-load stations at times of low demand (eg during the night) or by intermittently operating systems based on solar, wind or wave power (if these systems are eventually brought into operation). In such a scheme, known as a *hydroelectric pumped storage scheme*, the surplus electricity is used to drive motors which pump water from a low-level to a high-level reservoir. The pump motors are reversible so, when demand for electricity is high, the water is allowed to flow back down from the high-level to the low-level reservoir, generating electricity as it does so. Britain has four pumped storage schemes in operation or under construction, their total generating capacity being ~ 2800 MW.

Pumped storage schemes do not add to the total electricity supply; rather, they increase the efficiency of a generating system and provide system security (eg emergency stand-by capacity). The contribution of hydroelectric power to the UK's overall energy budget is therefore expected to remain at around its present level of roughly two mtce for the foreseeable future.

UK – Likely annual primary energy contribution by hydroelectric power by the year 2000: ~2 mtce

Geothermal power

Geothermal power stations make use of the earth's internal heat to produce steam that can then be used to drive turbines and generate electricity in a conventional manner. In many parts of the world, the temperature gradient near the surface of the earth's crust is too low to enable this to be done, but, in certain areas, known as *geothermal fields*, the temperature gradient is much higher. In some of these areas, steam can be produced simply by drilling to a depth of one to two km; in others, it is necessary to inject water which is then returned to the surface as steam. Figure 7.8 shows a schematic representation of the latter type of system.

Since geothermal power stations make use of steam to produce their electricity, the efficiency of conversion is effectively determined by the temperature and pressure of the steam used, as in a conventional power station. Although the energy is free, the steam conditions are nearly always such that the efficiency is fairly low. The total world electricity generating capacity from geothermal energy was roughly 1300 MW in 1973 – less than the output of a single large modern power station. The major 'geothermal powers' are the USA, Italy and New Zealand; Italy, for example, runs its electric train network from such power.

Geothermal sources can also be used for space heating and horticulture, and even for industry if local outlets are available. It seems unlikely that geothermal electricity will be able to make anything other than a minimal contribution to the energy budget of either the USA or the UK before the turn of the century, and it is therefore not included in our projections.

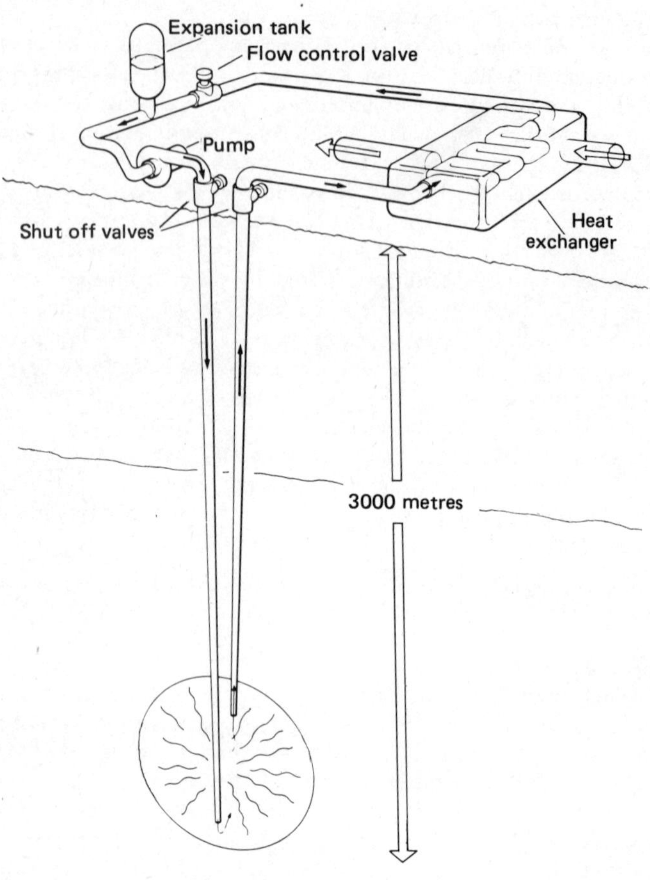

Figure 7.8 *A geothermal power scheme*

Will we have enough energy?

We are now in a position to draw up a table showing what the possible contribution from each of the resources examined above might be in the year 2000. It must be realized, however, that there is inevitably a considerable degree of uncertainty associated with all these resources. Some are uncertain owing to intrinsic resource limitations and/or political considerations, while others are uncertain because we have tried to assess the probability of technological and economic breakthroughs taking place. In line with our central philosophy, we have steered away from governmental predictions of how much energy we will need, and have also tried to assess the claims of the protagonists of the various alternative resources in a realistic manner.

In general, we feel that the USA will always be in a stronger position than the UK when it comes to obtaining imported supplies of oil, and the USA oil figure tries to reflect this position. We also feel that, despite the environmental problems surrounding strip mining of coal, coal production in the USA will increase at a much greater rate than in the UK. Conversely, the development of alternative resources will probably move faster in the UK than in the USA.

In the table below (Figure 7.9), we deliberately exclude any possible contributions from nuclear power, since the main object of this part of our book is to try to determine whether it would be possible to meet our future energy needs *by relying solely on non-nuclear energy sources.*

	Possible energy contribution in USA (mtce)	**Possible energy contribution in UK (mtce)**
Fossil fuels (non-renewable)		
Coal	1350	155
Oil	1350	100
Natural gas	300	40
Renewable energy resources		
Hydropower	175	2
Solar energy	60	10
Wind energy	20	15
Wave energy	—	25
Total	3255	347

Figure 7.9 *Likely energy supply pattern in the year 2000*

To put these figures into perspective, let us now compare them with the figures for likely total energy *demand* in the year 2000 that were obtained in Chapter 6 from our study of the historic growth, technifix and zero growth scenarios. These are tabulated in Figure 7.10, and are compared with the figures for likely total available non-nuclear energy in Figure 7.11.

	Likely total energy demand in 2000 USA (mtce)	**Likely total energy demand in 2000 UK (mtce)**
Historic growth scenario	4854	539
Technifix scenario	somewhere in range 2890-4854	somewhere in range 335-539
Zero growth scenario	2890	335

Figure 7.10 *Likely total energy demand in 2000 under different scenarios*

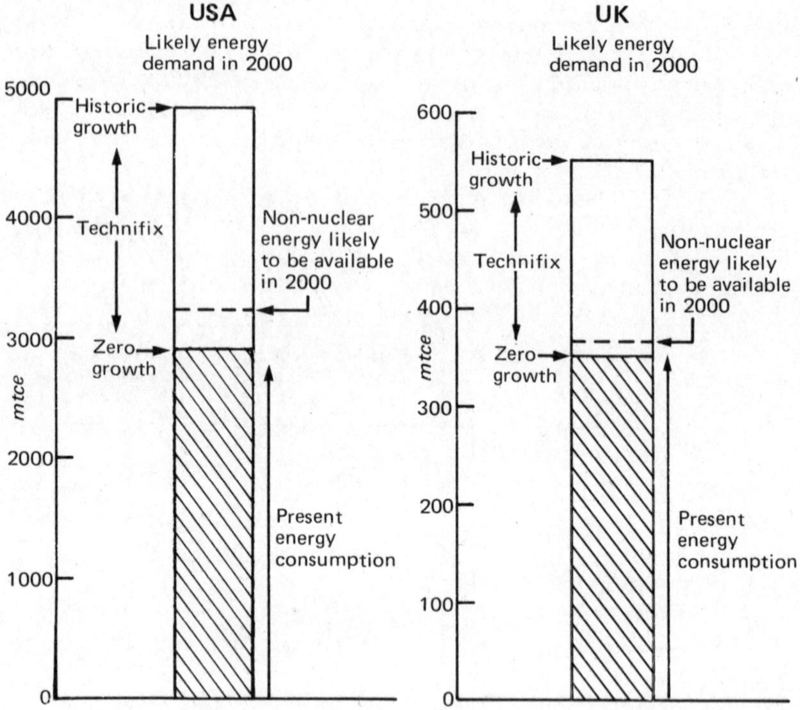

Figure 7.11 *Energy supply and demand in 2000*

Bearing in mind the fact that the likely energy contributions given in Figure 7.9 are, if anything, on the optimistic side, we see that there is no possible way in which either the USA or the UK could sustain the sort of growth in energy demand to which they have become accustomed during the last 30 years without making massive use of nuclear power. America seems to be slightly better off than Britain, and could (it would appear) sustain a modest increase in energy consumption without using nuclear power – at least up to the year 2000. In Britain, on the other hand, it seems highly unlikely that any significant increase in energy consumption will be possible without making extensive use of nuclear power.

The situation is likely to become even worse after the year 2000, since the contributions to the total energy budget from oil and natural gas seem certain to decline fairly rapidly during the first few decades of the 21st century unless massive new reserves can be found or technologies for the economic extraction of oil from shale can be developed. Thus, the need for nuclear power is likely to become progressively greater in both countries after the turn of the century, and we believe that the governments of Britain and the USA should be making plans *now* to ensure that the energy is available when it is needed. Just how they could do so will be the subject of Part III of this book.

PART THREE

A Possible Long-term Nuclear Strategy

The long-term future of nuclear power depends on several factors – whether or not an expansion of nuclear power is necessary, whether or not there are sufficient reserves of uranium available and whether or not nuclear power is socially and environmentally acceptable.

It has already been shown (in Part II) that, unless we are prepared to gamble on the discovery of massive new reserves of fossil fuels and on new, as yet unproven, technologies, we are likely to need a substantial nuclear power programme by the turn of the century. Whether or not nuclear power is socially and environmentally acceptable will be the subject of Part IV. In Part III, we will look at some of the options available for the future of nuclear power, and present the case for a mixed thermal/fast nuclear power programme. Such a programme could, we believe, sustain the industrialized nations of the Western world at least until the potential of fusion power has been thoroughly evaluated or the various alternative technologies have proved themselves to be practicable substitutes for our present sources of energy.

CHAPTER EIGHT

The History of Nuclear Power

In this chapter, we will review the first 25 years of nuclear power. After summarizing the events that led to the birth of commercial nuclear power programmes, we will look in more detail at the development of nuclear power in the USA, UK and France. These nations have always pursued independent policies with regard to nuclear power, and still adopt radically different postures on its future role. The comparisons should prove useful, however, because they will lead us on naturally to the discussion of possible future nuclear power options that will be given in Chapter 9.

The beginnings of nuclear power

Although many people think that the nuclear age began on August 7 1945, it actually began several years before this with the discovery of nuclear fission in 1937 by Hahn, Strassmann and Meitner. (Strictly speaking, Fermi and his co-workers had already achieved fission in 1934, but had failed to appreciate the significance of their results.) The first man-made nuclear reactor was constructed at the University of Chicago by Enrico Fermi and his colleagues. It went critical in 1942, and became known as the 'Fermi Pile'. The reactor demonstrated the feasibility of producing controlled nuclear fission only a little more than four years after the discovery of the fission process itself, and, although the Fermi Pile led directly to the development of the Hiroshima and Nagasaki bombs, it was also the first vital step towards the peaceful application of nuclear power.

The credit for building the world's first commercial nuclear power station is normally given to Britain, but it can be argued that the Russians have a strong prior claim. The first nuclear-powered electricity generating plant to drive a dynamo was in fact built by the Russians at a station near Moscow, and was commissioned in 1954. This was a low-power station producing only a few megawatts, however, and was probably only a demonstration reactor system rather than a fully commercial plant. Three years earlier, the Americans had produced electricity from a research reactor at the Argonne National Laboratory, but it was not until 1957 that their first fully commercial nuclear power station was put into service at Shippingport, Pennsylvania.

In the UK, a team at the Government Research Establishment at

Harwell in Berkshire had completed a design study for what was eventually to become Calder Hall Nuclear Power Station as early as 1950. The main purpose of this work was to develop reactors capable of producing weapon-grade plutonium more efficiently than those currently in use at Windscale, but it was also intended to use the heat generated by the reactors to produce electricity. Calder Hall was switched on by Queen Elizabeth on 17 October 1956.

By the late 1950s and early 1960s, the construction of civil nuclear power stations was well under way. By now, however, the most powerful advocates of the expansion of civil nuclear power programmes were no longer the research and development scientists but the political figures of the day. As a result, nuclear power flourished.

Nuclear power in the USA

In 1954, the Atomic Energy Act, which permitted private ownership of nuclear reactors and the private use of nuclear fuels, marked the beginning of the American civil nuclear power programme. The design of the original Shippingport reactor was based on that used in the nuclear submarine 'USS Nautilus' in 1954. This was a pressurized water reactor, and, as was explained in Chapter 2, American interest in water reactors for power generation can largely be attributed to the success of their early work on such reactors for marine and submarine propulsion.

Under the 1954 Act, Electricity Utilities were encouraged to join the national effort to prove the technical and commercial viability of fission-powered generation systems. The Cooperative Power Demonstration Reactor Programme was responsible for the development and testing of several different designs of reactor plants, including an experimental boiling water reactor, a pressurized water reactor, a sodium-cooled graphite reactor, an experimental breeder reactor and an experimental homogeneous reactor. By 1963, however, the major reactor manufacturers had begun to commit their production lines to light water reactors, particularly the boiling water reactor (BWR) and the pressurized water reactor (PWR).

Throughout the above development programme, considerable attention was also given to fast reactor work (remember that the first 'nuclear' electricity was produced from a fast reactor – see Chapter 2). Two such reactors were operating by the mid-1960s, but one was eventually converted to a fuel testing facility (which is still operating as such) and the second, following a coolant flow failure and sub-assembly meltdown, was ultimately decommissioned. (Major plant redesign had been considered necessary to bring the system within required safety criteria, and the projected high costs of the changes were cited by the private owners as their reason for discontinuing the project.)

One of the chief characteristics of the American nuclear programme was its emphasis on the use of enriched uranium, as opposed to natural uranium. Extensive research had indicated that use of the former would lead to improved performance and cheaper plants, and, in addition, a plentiful supply was available from American diffusion plants.

Utility orders for nuclear reactors increased sharply in 1966, and, by the beginning of 1968, when the Utilities were operating a total of only 1500 MW of plant, construction had begun (or orders had been placed) for no fewer than 62 reactors with an additional generating capacity of roughly 50,000 MW. This rate of ordering was, of course, prompted by the phenomenal growth in electricity demand during the 1960s (see Chapter 5), which indicated that generating capacity would again have to be expanded during the 1970s. The size of nuclear power stations also gradually increased during this period. In 1965, the average size was roughly 650 MW, but, by 1969, this had increased to over 1000 MW. At a time when cheap oil was plentiful, this increase in unit size did not bring immediate economic benefits, however, since average unit capital costs were escalating and construction periods and licensing periods were also increasing.

By the end of 1978, the total operating generating capacity of nuclear power stations in the USA was roughly 52,000 MW. There were 71 operational units, which accounted for over 12 per cent of all electricity generated; in addition, 73 reactors were under construction (79,000 MW) and a further 19 had plans approved.

At the time of writing, the Carter moratorium on fuel reprocessing (1977) and the incident at the Three Mile Island nuclear power plant (March 28 1979) have combined to bring the American civil nuclear power construction programme to a virtual halt. Work is continuing on fast reactor development, however, and the level of funding for this work is still significantly larger than that for the whole of Western Europe, although there are no firm plans for the commercialization of fast reactors as yet.

Nuclear power in Britain

In contrast to the American programme, the development of nuclear power in Britain has concentrated almost entirely on gas-cooled reactors. It has, however, been realized for many years now that the fast reactor represents an enormous untapped potential for conserving fuel, and, as a result, Britain has almost 30 years of fast reactor experience to build upon.

The British thermal nuclear power programme to date falls neatly into four distinct phases.

The Magnox programme

During the 1950s, a full-scale commercial nuclear power programme was initiated with the ordering of 11 power stations using reactors of the Magnox type (see Chapter 2). The first two stations (Calder Hall and Chapelcross) were built by the United Kingdom Atomic Energy Authority (UKAEA) and the remaining nine by the Central Electricity Generating Board and the South of Scotland Electricity Board. Calder Hall (now operated by British Nuclear Fuels Ltd) is shown in Figure 8.1. The principal design features of Britain's various Magnox stations are given in Figure 8.2.

Figure 8.1 *An aerial view of the Calder Hall nuclear power station*

Station	Current operator	Number of reactors	Design electrical output per reactor (MW)	Typical station output (MW)	Date of commissioning
Calder Hall	BNFL	4	50	198	1956
Chapelcross	BNFL	4	50	186	1958
Berkeley	CEGB	2	138	276	1962
Bradwell	CEGB	2	150	250	1962
Hunterston A	SSEB	2	160	300	1964
Hinckley Point A	CEGB	2	250	430	1965
Trawsfynydd	CEGB	2	250	390	1965
Dungeness A	CEGB	2	275	410	1965
Sizewell A	CEGB	2	290	420	1966
Oldbury	CEGB	2	300	416	1967
Wylfa	CEGB	2	590	840	1971

Figure 8.2 *Britain's Magnox nuclear power stations*

It was originally intended that the Magnox programme would be completed by 1968 and that it would contribute a total generating capacity of 5000 MW. In practice, however, the programme was not completed until 1971, and, in addition, most of the stations had to be down-rated by roughly ten per cent in order to avoid possible fatigue cracking and corrosion in the mild steel of which their steam systems were constructed. Thus, the actual generating capacity eventually achieved by the Magnox programme was less than 4500 MW.

Britain's Magnox stations have proved to be extremely reliable, and are expected to have operational lives of up to 30 years – ten years longer than their original design life. Apart from her various hydroelectric power stations, they are now Britain's cheapest source of electricity. The Magnox reactor is also a highly efficient source of plutonium, producing more plutonium per unit of electricial output than the CANDU, PWR or AGR – or even the liquid metal fast reactor. To date, over 13 tonnes of plutonium and 20,000 tonnes of depleted uranium have been produced as by-products of the Magnox fuel cycle.

The AGR programme

During the late 1950s, Britain started work on a more ambitious reactor design – the AGR (see Chapter 2). An experimental prototype of 30-MW generating capacity was commissioned in 1962. As a result of early operating experience with this small prototype, it was considered that the engineering features of the new reactor design were sufficiently close to those of the Magnox reactors then being built to obviate the need for the construction of a large-scale prototype. A series of five full-scale commercial AGR stations were therefore ordered in 1964, designed to provide an extra six GW of generating capacity by 1975.

Events have subsequently shown that the decision not to build a large-scale prototype was a serious error of judgement. The programme is badly behind schedule, mainly owing to unexpected problems with the steam-generating plant and the turbines (not, it should be stressed, with the reactors themselves). By 1977, only 1500 MW of the total planned AGR capacity was available, and all AGR stations have now been down-rated. It is now expected that the total generating capacity of the five original AGR stations will be less than five GW when they are fully operational – by 1982, it is hoped.

So far, no AGR fuel has been processed on a commercial scale, but a reprocessing plant capable of handling the ceramic oxide fuel used in these reactors is currently under construction at Windscale (see Chapter 3).

The SGHWR programme

Britain's first venture into water reactors took place in the 1960s with the development of the Steam Generating Heavy Water Reactor. (Previously, it had been felt that gas-cooled reactors had a very high degree of intrinsic safety, exposed the operators to very low radiation, and produced low radioactive discharge compared with water reactors.) A prototype station was constructed at Winfrith in Dorset, becoming operational in 1967 and, in 1974, two full-scale commercial stations were planned. These would have had a combined generating capacity of four GW, but were cancelled in 1978 before construction began.

There were several reasons for abandoning the SGHWR programme. Firstly, the fall in demand for electricity brought about by the 1974 Middle East oil crisis meant that Britain would not need nearly as much new plant as had previously been predicted. Secondly, progress with the detailed design of a commercial SGHWR had been disappointing, largely because of attempts to produce operator radiation levels and other safety features to a higher standard than could be offered by other water reactors. Because the AGR programme was now beginning to produce results, it was decided instead to build two further AGR stations – at Heysham (CEGB) and Torness (SSEB). These would come on stream at a time when it would be necessary to start thinking seriously about replacing some of the old, uneconomic fossil-fuelled stations, and perhaps some of the early Magnox stations as well.

The PWR programme

Britain seems to have had little success in selling its reactor designs abroad. To date, only two reactor systems have been exported – one to Italy and one to Japan. (These were both of the Magnox design.) Water reactors currently dominate the world scene, and, although it was hoped that the AGR system would prove an attractive alternative, this has not so far proved to be the case. The failure to develop a successful water reactor capable of competing with American designs on the world market is a major cause for concern among British reactor manufacturers.

In 1980, plans were announced to build a station based on the American PWR system. This is to be built at Sizewell on the East Anglia coast, and the decision to import reactor technology represents a major policy reversal on the part of the British nuclear power industry.

Planning for the future

At present, the intention of the British government is to order new nuclear power stations on a regular basis over the next decade, so that Britain will have roughly 20 GW of operational nuclear capacity by the turn of the century. At the time of writing, it is not clear what type of reactors (or how many) will be ordered during the late 1980s, since this will depend to a large extent on the safety evaluation currently being carried out on the PWR system, and also on the way in which electricity demand moves in the next few years. It seems, however, that Britain will almost certainly build no more Magnox stations, and that any new stations will probably use the AGR and/or PWR designs.

The UK fast reactor programme

In 1954, the British government authorized the construction of an experimental power-producing fast reactor at Dounreay – on the northernmost tip of Scotland, near Thurso, in Caithness. This reactor, which became known as the Dounreay Fast Reactor (DFR), was to be used to ascertain whether such reactors could be operated safely and whether they were likely to be able to produce electricity at a reasonable price. Work started on DFR in 1955, and was completed in 1959, when it became the first fast reactor in the world to supply electricity to the general public through a national grid. DFR's electrical output was 15 MW. The reactor was also unique in that it was supported by a total on-site fuel cycle, all its fuel being manufactured and reprocessed at Dounreay.

By 1963, DFR's original proving programme had been so successful that its role was changed to that of a test-bed for fast reactor materials and components, and, during the next 14 years, it earned valuable consultancy contracts from a large number of overseas customers. During its operational lifetime, DFR generated over 600 million units (kWh) of electricity. It was finally closed down in 1977, but the famous Dounreay sphere will probably not disappear from the landscape. It is intended that, after full decommissioning, it will be used to house a fast reactor exhibition centre.

By 1966, the British government had decided to extend its fast reactor programme, and approval was given for the construction of a much larger fast reactor at Dounreay – the Prototype Fast Reactor (PFR). PFR, which was completed in 1974, has an output of 250 MW, and is proving to be an exceptionally safe, reliable and efficient reactor design. Since it is a true prototype for a future commercial fast reactor, most of its work is devoted to the testing of components and systems for such a reactor under actual operating conditions. The original reprocessing plant at Dounreay, which was built to handle DFR fuel, has been decommissioned and converted to handle PFR fuel. It successfully reprocessed its first batch of PFR fuel early in 1981.

The question of whether or not Britain should follow the lead of France and the USSR and order a commercial demonstration fast reactor (CDFR) is still under consideration. For reasons which are explained in the next chapter, the authors hope that the construction of a CDFR will be approved before the end of the 1980s.

The present status of the British nuclear power programme is summarized in Figure 8.3, which shows the disposition of all existing and planned nuclear stations.

Nuclear power in France

The French nuclear power programme contains elements of what has been happening in the USA and Britain. It began in the early 1950s with a programme of nine gas-cooled graphite reactors, which were built between 1954 and 1972. This parallels the Magnox developments in the UK.

In 1967, however, France's first Westinghouse-type PWR plant was commissioned, and this 280-MW station proved to be the forerunner of the current French thermal reactor programme. In 1970 Electricité de France (EDF) made the decision to place 50 per cent of its future investment into nuclear power plants; in 1973, the nuclear commitment was increased to two thirds and, in 1974, it was decided that nearly *all* new power plants would be nuclear. By 1985, it is expected that over 50 per cent of French electrical power will be generated by nuclear stations.

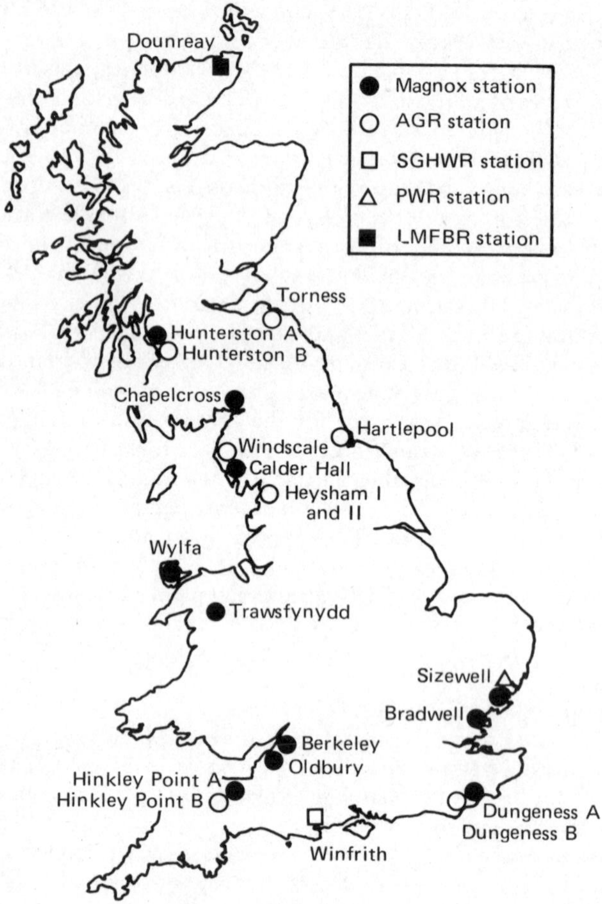

Figure 8.3 *Britain's nuclear power stations*

This ambitious programme, which is in stark contrast to the situation in the USA, is based on over 40 stations, of which 20 are already operational and the remainder are under construction. By 1985, it is planned to have 35 GW of generating capacity from PWR stations alone.

The French are also the leaders in the West when it comes to fast reactors. Their Super Phénix, which will have a generating capacity of 1200 MW, is expected to be operational by 1983. EDF has also announced its firm intention to order two 1500-MW fast reactors before 1985.

The intention is quite clear: the Commissariat à l'Energie Atomique (CEA) feels it has a responsibility to the people of France, who have put their faith in nuclear power, to develop an energy system which will greatly reduce their country's dependence on oil. In many ways, the

French are now leading the world in the field of nuclear power, and they intend to stay in that position.*

World growth of nuclear power

The growth of nuclear power throughout the world since 1960 can only be described as phenomenal. In the decade to 1970, the number of installed reactors increased from 24 to 98, and total generating capacity went from 1300 MW to 20,000 MW – an increase averaging *31 per cent* per annum. By the end of 1980, the number of reactors had risen to 238, and the installed generating capacity stood at 124,500 MW. The rate of growth in this last decade was lower – about 20 per cent per annum – but there are still a large number of new systems either under construction or being planned, as can be seen in Figure 8.4 below.

	Number of reactors	Generating capacity (MW)
Operating reactors	238	124,493
Reactors under construction	230	211,359
Reactors planned	111	103,554
Total	579	439,406

Source: IAEA

Figure 8.4 *Current status of the world nuclear power programme (1980)*

By 1980, the contribution of nuclear power to the production of electricity throughout the world had reached roughly eight per cent. This is expected to rise to about 12 per cent in 1985, when the installed capacity will be approaching 350,000 MW – equivalent to the oil production of Saudi Arabia in 1978. Of the 579 reactors mentioned in Figure 8.4, 323 are of the PWR design, with BWRs and gas-cooled reactors making up a further 248. Thirteen fast reactors are also included; indeed, 1980 saw the commissioning of the world's largest fast reactor in Russia (600 MW).

As the various reactors currently being constructed or planned are brought into operation, the demand for supplies of natural uranium will begin to increase rapidly, and it is now being questioned whether such a growth rate can be sustained for many more years. In the next chapter we will look at this aspect of nuclear power in more detail.

* The above was written before France's 1981 Presidential Election. The new Socialist President, M Mitterand, is much less pro-nuclear than his predecessor, and this may well affect France's long-term plans.

CHAPTER NINE

A Possible Nuclear Power Strategy for the Western World

Having established in Part II that significant amounts of nuclear power are likely to be needed in order to avert a severe energy shortage early in the 21st century, we will now try to determine what type of nuclear programme would best meet our future energy requirements.

We will begin our discussion by taking a brief look at the economics of nuclear power -- something that we have not done so far. Using Britain as a case study, we will compare the costs of nuclear and non-nuclear electricity, and will then show how the British cost figures compare with those of other important nuclear nations.

Next, we will examine the feasibility of continuing or expanding our existing thermal nuclear power programme into the 21st century, and will show how this would inevitably lead to the rapid exhaustion of the world's supplies of economically recoverable uranium.

Finally, we will discuss the potential role of the fast reactor, and will show how it would be possible to extend the effective life of the world's uranium reserves by embarking on a 'mixed' nuclear power programme – building up our fast reactor capacity with the aid of plutonium produced in thermal reactors until the latter can eventually be completely replaced by self-sustaining fast reactors.

The economics of nuclear power

Apart from hydroelectric power, nuclear power now represents by far the cheapest way of generating electricity. Nuclear power stations are, and always have been, expensive to build, the capital cost per installed kW of generating capacity being up to three times that of a coal-fired or oil-fired station. The running costs of nuclear power stations are, however, very much lower, owing to the large fuel cost advantage that such stations have over fossil-fuelled plants. Because of these low running costs, the overall cost of producing electricity from a nuclear power station (taking into account all operating costs, fuel costs, repayments of capital cost, interest charges, etc) is, in fact, now significantly lower than for coal or oil stations.

When we started to build nuclear power stations during the 1950s and 1960s, it was always accepted that, in competition with the (then) cheap coal and oil and the well-established technology of fossil-fuelled stations, the cost of nuclear electricity would not be competitive before the 1970s at the earliest. The licensing and safety regulations that nuclear stations had to comply with were always very much more stringent than for fossil-fuelled stations, and this frequently resulted in high levels of capital cost escalation. It can, however, be argued that fossil fuel stations are now experiencing the same sort of environmental pressures, and that nuclear stations are nearer to the end of such expenditure. In addition, the fuel costs of fossil fuel stations have risen dramatically in recent years, so that most commentators now agree that nuclear electricity is, and will remain, cheaper than non-nuclear electricity.

Another measure of the economic benefit of any power-generating system is what is termed the *net effective cost*. The net effective cost of nuclear power is in fact now considered to be *negative*. This means that the capital cost of building a new nuclear power station can now be more than recovered over its working life by the net savings in overall running costs that the introduction of the station makes possible. New, highly efficient coal stations can also produce substantial system cost savings, but, with increasing fuel costs, these savings are not likely to be large enough to cover the capital cost of the new plant. Typical figures produced by the CEGB in the UK show that, for stations of a similar size, a new nuclear station will have a net effective cost of – 0.46p per unit (kWh) compared with +0.33p per unit for a coal-fired station (1980 figures).

Britain's original Magnox stations have already resulted in total savings of over £100 million compared with the coal-fired alternative, although the programme has not yet achieved a negative net effective cost.

Some members of the anti-nuclear lobby claim that nuclear electricity is *still* more expensive than that from fossil-fuelled stations, and accuse the Electricity Boards of hiding the true costs from the general public. They say that the true cost of producing electricity from new nuclear power stations is probably about 25 per cent above that of coal- and oil-burning plants, and that the true figures are distorted by such devices as nominal interest rates, selective reactor costings and abnormal accounting methods. These claims are, of course, strongly denied by the Electricity Boards concerned, and appear to be unsustainable in the light of the evidence. The Central Electricity Generating Board (UK) published its 1979-80 Annual Report in July 1980, and we produce two comparative tables in Figures 9.1 and 9.2. (Details of the full report are given in the Bibliography.)

In Figure 9.1, the costs per unit (kWh) are given for all types of stations commissioned during the period 1965-77, and are based on the following cost headings: capital charges and provision for decommissioning, interest during construction, inclusive fuel costs, other operating costs, and research and training.

In Figure 9.2, the CEGB has estimated similar costs for stations that are currently under construction; in this case, the full costs are based on the estimated real costs at the time of commissioning. This new way of presenting cost breakdowns should go a long way towards answering the

Year of operation	Cost per unit produced by nuclear (Magnox) stations	Cost per unit produced by coal-fired stations	Cost per unit produced by oil-fired stations
1978-79	1.12p	1.33p	1.35p
1979-80	1.30p	1.56p	1.93p

Figure 9.1 *Comparative costs per unit of electricity generated by stations commissioned in the period 1965-77 (CEGB)*

anti-nuclear arguments about governments and generating boards hiding the true figures from the general public. (As with other aspects of nuclear power, the authorities have, until recently, been unwilling to publish detailed information in a form that could be understood by people outside the industry. For this new manifestation of 'open government', we can thank pressure groups like those that make up the anti-nuclear lobby, for it is only as a result of their efforts that information of this type is now starting to be made available.)

Nuclear stations (AGR)			Coal-fired station	Oil-fired stations	
Dungeness (B)	*Hartlepool*	*Heysham (1)*	*Drax (2)*	*Littlebrook B (D)*	*Ince (B)*
2.62p	2.28p	2.31p	3.59p	6.63p	7.14p

Figure 9.2 *Estimated comparative costs per unit of electricity generated for stations currently under construction (CEGB – 1980 prices)*

The summary figures in Figure 9.2 tell their own tale; it would obviously be economically disastrous to order more oil-fired plant, and new nuclear plant is seen to be even more competitive with coal-fired plant that it was a decade ago.

A similar pattern of comparative costs between nuclear and coal-fired stations has emerged in the USA over the last decade. By 1974, for example, American nuclear stations were producing electricity at a total cost that was roughly ten per cent lower than that from coal-fired stations.

France, with its ambitious nuclear power programme, seems to have come to a similar conclusion, as the figures given below show. These relate to generating costs at 1979 fuel prices.

Coal-fired stations – 2.3p/kWh
Oil-fired stations – 2.5p/kWh
PWR stations – 1.4p/kWh

In fact, after an exhaustive examination of the available data from a large number of countries, *we can find no cost analysis which shows that electricity generated in fully commercial nuclear power stations is more expensive than that from equivalent coal- or oil-fired stations.*

Because of the rapidly increasing costs of fossil fuels, nuclear power is expected to become even more attractive from an economic point of view in the short term. The prices of coal and oil are both expected to rise by at least five per cent *in real terms* throughout the 1980s, whereas the operating costs of nuclear power stations are not expected to rise at anything like this rate. Thus, since the capital costs of coal, oil and nuclear stations are all expected to rise at roughly the same rate (more or less in line with inflation), the cost of producing a unit of electricity from a fossil-fuelled station is likely to become progressively higher than the cost of producing a unit of electricity from a nuclear-powered station (Figures 9.1 and 9.2 show this trend clearly).

The development of alternative sources of energy is also unlikely to provide a serious economic challenge to nuclear power in the short term, since (a) such sources are unlikely to become available on a large scale before the end of the century, and (b) electricity produced from such sources will most probably be more expensive than nuclear electricity for some time to come. Most commentators believe that the provision of base-load electricity from alternative sources of energy is still a long way off.

Because of the undoubted economic advantages of nuclear power over other methods of generating electricity, it would be very nice if we had the option of expanding our thermal nuclear power programmes as our supplies of oil and natural gas become exhausted, thus guaranteeing a cheap source of base-load electricity throughout the 21st century and beyond. There is, however, a cloud on the horizon, namely the limited nature of the world's uranium reserves. Let us now see how this is likely to affect the situation.

World uranium supplies and limits to the use of thermal reactors

Because the life of a nuclear power station can be up to 30 years, it is absolutely necessary to be able to guarantee that fuel will be available for the station at a reasonable cost for such a period. If this cannot be done, then the economic viability of the station must be in doubt. We have already seen what happened to the economic competitiveness of the oil-fired stations that were planned in the late 1960s and early 1980s, before the 1974 Middle East crisis. Although fuel costs are relatively less important in the case of nuclear power stations, a dramatic rise in the cost of uranium could influence the overall economics in an undesirable fashion, and, if the costs became too great, could make the situation uncompetitive.

Uranium is not a particularly rare metal, constituting roughly 2.8 parts per million (by weight) of the earth's crust. Indeed, the top one km of of crust is estimated to contain roughly one million million tonnes of uranium. In practice, however, it is very much scarcer than these figures would appear to suggest, since virtually all the above uranium is present in such low concentrations that it cannot be recovered at an economic rate. It is generally agreed that the present recoverable reserves (or *reasonably assured reserves* as they are sometimes called) of uranium in the non-communist world are roughly 1.8 million tonnes, a figure that is increased to roughly five million tonnes if we include (a) lower-grade ores that *might* become economic if the real price of uranium rises, and (b) new reserves that are expected to be discovered during the next few decades. These various estimates of our reserves are summarized in Figure 9.3.

Extraction cost (1980 prices)	Reserves (millions of tonnes)
below £35 per kg	1.8
£35–£55 per kg	0.7
above £55 per kg and new reserves	2.5
Total	5.0

Figure 9.3 *Uranium reserves — non-communist world*

At the moment, the non-communist world has roughly 120,000 MW of nuclear generating capacity, and consumes natural uranium at a rate of roughly 30,000 tonnes per annum. As we saw in the last chapter, the non-communist world's total nuclear generating capacity is expected to increase steadily during the 1980s, and, by 1990, will probably be somewhere between 370,000 MW and 460,000 MW, depending on the rate at which the world economy expands. At present consumption levels, the recoverable reserves would last for at least 50 years, but, by 1990, the annual consumption rate will have increased to between 60,000 and 80,000 tonnes. By 2000, it is expected that consumption could be as high as 140,000 to 200,000 tonnes per annum. Thus, between 1980 and 2000, it can be calculated that the non-communist world will probably use up somewhere between 450,000 and 800,000 tonnes of uranium, leaving between one million and 1.35 million tonnes of the total estimated recoverable reserves of 1.8 million tonnes. At the consumption rate expected by the year 2000, the life of these remaining reserves would be five to ten years at most. Even assuming that all of the five million tonnes listed in Figure 9.3 can ultimately be classed as 'recoverable' the reserves will only last until the third or fourth decade of the next century at the latest.

Thus, we seem almost certain to be confronted with a world shortage of uranium early in the 21st century — when the new thermal stations now being planned will still be in operation. Because of the scale of the proposed expansion in nuclear power, it is also conceivable that there may be difficulties in providing uranium at the required rate, particularly if most nations decided to adopt the *once-through fuel cycle* that the Americans and Canadians are now using.

In Chapter 3, the various possible nuclear fuel cycles were described in some detail, and it was explanined that the raw ore requirements of any thermal nuclear power programme are strongly dependent on the type of fuel cycle that is adopted. If the unused uranium and/or the recovered plutonium are recycled into the reactors, the total requirement for fresh supplies of uranium ore can be reduced.

We also mentioned, at the beginning of Chapter 3, that the International Fuel Cycle Evaluation Exercise (INFCE) was launched in 1977 in order to assess the international implications of the various nuclear fuel cycles. Since INFCE was, in effect, a direct response to President Carter's moratorium on fuel reprocessing and the commercial recycling of plutonium, its conclusions and recommendations relate closely to the proliferation issue. These will be discussed in more detail in Chapter 12, but, for the moment, it can be said that INFCE considered the problems of proliferation to be basically political rather than technological in nature. With regard to assuring supplies of nuclear fuels, for instance, INFCE felt strongly that any uncertainty over the availability of nuclear materials or existing fuel cycle services could lead countries to seek independence by setting up their own fuel cycles. This could then lead to premature decisions to construct reprocessing plants.

INFCE recognized that there are political and environmental problems which limit the exploitation of uranium, and urged that efforts should be made to find acceptable solutions. INFCE also recommended that reactor strategies which *conserve* uranium should be adopted by the major users of nuclear power. As we have seen in Chapter 3, there are several ways in which this can be done. Use of the *reprocessing cycle*, which recovers the unused uranium and recycles it to the enrichment plant (see Figure 3.5) can reduce the 'raw' uranium requirements of most thermal reactor systems by over 20 per cent. In the *thermal recycle fuel cycle*, the plutonium that is recovered during reprocessing of the fuel is also fed back into the thermal programme as a component of the reactor fuel. Although this reduces the raw uranium requirements by a further 16 per cent, INFCE found that the economic advantages of such a process were slight, and concluded that most nations would probably prefer to look to the fast reactor as a means of making the best economic use of plutonium.

How the use of fast reactors could improve the uranium supply situation

There are two strong arguments in favour of the commercial development of fast reactor systems as opposed to the purely thermal systems currently in use throughout the world. Firstly, the use of fast reactors could enable up to 60 per cent of the energy contained in natural uranium to be released, as compared with only one to two per cent in a purely thermal programme. Clearly, this would lead to a much more efficient use of the world's supplies of natural uranium, and could greatly extend the life of our reserves. Secondly, fast reactors represent the only peaceful method of destroying plutonium, since they can, by suitable design of their cores, be used to consume more plutonium than they create rather than create more than they consume, as they do when used in a breeder role. This could be important if it were ever decided to run down world stocks of plutonium.

Let us now examine the fuel consumption characteristics of fast reactors in greater detail.

The relative efficiencies of fast and thermal reactors

In terms of actual consumption of fissionable material for a given amount of electricity generated, different reactor types (thermal and fast) seem to behave in a similar way, any differences being due to the efficiencies of the steam cycles used rather than to the characteristics of the reactors themselves. A comparative table which demonstrates this effect is shown in Figure 9.4. Note that the amount of fissile material needed to produce a given amount of heat is virtually independent of the type of reactor.

Reactor type	AGR	PWR	CFR
Fissionable mass consumed in producing 1 gigawatt year of heat (kg)	380	380	383
Overall station efficiency	41%	32%	39%
Fissionable mass consumed in producing 1 gigawatt year of electricity (kg)	926	1186	982

Figure 9.4 *Fissionable mass requirements for equivalent power output in different reactor designs*

In the AGR and PWR reactors, the fissionable material is enriched uranium, where the U235 content is between two and three per cent as opposed to the natural level of only 0.7 per cent. To make one tonne of enriched uranium at an enrichment level of, say, 2.5 per cent requires 5.5 tonnes of natural uranium, however, the remaining 4.5 tonnes being rejected from the enrichment process as depleted uranium contain only ~ 0.3 per cent U235. The enrichment process constitutes the main inefficiency in the use of uranium in thermal reactors, because the vast majority of the natural uranium used ends up as depleted uranium which is of no further use for thermal reactors. There are also losses of uranium in the chemical plants associated with ore preparation and reprocessing, but these are small in comparison to the enrichment losses.

To run a reactor obviously requires much more fuel than the one tonne that actually undergoes fission. Burn-up factors are nothing like 100 per cent, and so the one tonne of fissioned material has to be supported by a further *40-50 tonnes* of fuel which does not undergo fission. Thus, in order to run a typical thermal reactor power station in which roughly one tonne of fuel is fissioned in one year, it is necessary to mine *up to 250 tonnes* of natural uranium. Even though, as we have seen, reprocessing can result in significant savings, the annual natural uranium requirement of a thermal station is unlikely ever to be less than ~100 tonnes. More detailed figures have been produced for particular reactor designs and operational modes, but these are not really any more informative than the general case examined above. The differences between the AGR and PWR, for example, are slight provided that they operate in the same fuel cycle.

With the fast reactor, which burns plutonium and can produce (if required) more fissile material than it consumes, there need be no wastes from enrichment plants. Other process wastes can be calculated on the same basis as for thermal reactors, and these constitute the principal losses in the fast reactor fuel cycle. On this basis, an AGR or PWR station which operates on the reprocessing fuel cycle and requires a natural uranium input of ~150 tonnes per year can be compared to a fast reactor station of similar generating capacity which requires a natural uranium input of *only 2.5 tonnes* per year.

From this typical example, we see that the real saving in natural uranium consumption could be extremely large if fast reactors were exploited commercially. If the price of uranium increases (as seems likely if the current expansion of thermal nuclear power programmes continues worldwide) then the case for the fast reactor becomes even stronger. Unfortunately, it is not simply a matter of deciding to build fast reactors instead of thermal reactors. Even if the political and social implications of fast reactors were to be accepted without further discussion, the development of a viable fast reactor programme would be critically dependent on the availability of plutonium – a material that can (initially at any rate) only come from the reprocessing of spent fuel from a *thermal* nuclear programme. We can demonstrate this point most effectively by looking at Britain and the USA.

Britain and the fast reactor

It was estimated that, in 1980, there were roughly 13 tonnes of plutonium in Britain. Of this total, only four tonnes had been recovered, purified and stockpiled, however, the remainder being contained in fuel elements that were still in store prior to reprocessing (~ one tonne) and in the cores of thermal reactors (~ eight tonnes).

Each 1000 MW of fast reactor generating capacity requires a total plutonium inventory of up to four tonnes: ie sufficient to provide the initial fuel charge, and to provide replacement fuel until such time as the reactor becomes self-sustaining. From these figures, it is clear that Britain could only fuel *one* fast reactor at the moment.

By the end of the century, the combined Magnox, AGR and PWR programmes in the UK should have produced a further 35-45 tonnes of plutonium, and, when the oxide reprocessing plant at Windscale is in operation, this will allow more of the plutonium to become 'available' for use in fast reactors. Present plans indicate that the thermal nuclear programme will stabilize at around 22,000-MW capacity by the year 2000, and this should produce a further three to four tonnes of plutonium each year thereafter. Thus, by the end of the first decade of the 21st century, Britain should have enough plutonium available to provide the fuel for just over 20,000 MW of fast reactor generating capacity.*

This argument is, of course, based on the assumption that reprocessing of Magnox reactor fuel will continue in Britain, and that the oxide fuel from the AGR programme (and from the PWR programme, if there is to be one) will also be reprocessed. If, for whatever reason, Britain decided to halt its reprocessing operations, a fast reactor programme would not be possible.

The USA and the fast reactor

The thermal nuclear programme in the USA is roughly eight times the size of the British programme. Sufficient plutonium has already been recovered from the LWR programme to provide the full fuel inventory for several fast reactors. However, since 1977, the reprocessing of thermal reactor fuel has been halted, and spent fuel elements are now simply being stored. Whether this moratorium will be temporary or permanent is not clear at the time of writing. This once-through fuel cycle is particularly wasteful of natural uranium *and* of the plutonium that is produced as a by-product. Unless the American government is prepared to allow the reprocessing of thermal reactor fuel once again, the future of fast reactors in the USA is not very bright – despite the large amounts of money currently being spent on research and development in this area.

A possible mixed thermal and fast reactor programme

In view of the likely shortage of uranium supplies in the second quarter of the next century, the fast reactor seems to occupy a potentially vital position in any future energy strategy. Since it is clearly impossible to introduce an extensive fast reactor programme in the short term, we believe that the major nuclear powers should give serious consideration to the development of a *mixed thermal and fast reactor programme*, since this could go a long way towards alleviating this projected uranium shortage.

Such a programme would conserve large amounts of natural uranium as it was gradually introduced, and would remove progressively larger amounts of plutonium from stocks or from storage and transfer them to reprocessing plants and reactors. Perhaps even more significantly, the programme would give us the opportunity to make use of some of our by now huge stocks of depleted uranium. In the UK, for instance, there are

* These figures are based on the fact that Magnox, AGR and PWR reactors respectively produce roughly 620 kg, 180 kg and 270 kg of plutonium (net) for every 1000-MW year of continuous operation.

now over 20,000 tonnes of stockpiled depleted uranium (see Figure 9.5) and, if these could be used in a fast reactor programme, they would be equivalent (in energy terms) to the UK's present coal reserves. Put another way, this means that they could be used to meet *all* the UK's electricity needs for the next *300 years*. The USA has even larger stocks of depleted uranium, amounting to over 200,000 tonnes, and this figure could rise significantly if the current moratorium on reprocessing were to be removed. The potential for fast reactor development in the USA is therefore enormous.

In a mixed nuclear power programme, the number of fast reactors would initially be small, owing to the limited amount of plutonium available, and the rate at which such reactors could be built would be largely determined by the type of reprocessing cycle that was adopted and by the breeding rate that was employed in the fast reactors. Current designs of fast reactors can produce roughly 200 kg of plutonium more than they consume for every 1000-MW year of operation. With present load factors, this means that a fast reactor should, over its operational life, be able to produce enough plutonium to provide fuel for its replacement. *The initiation and subsequent expansion of a fast reactor programme is thus totally dependent on the existence of a parallel thermal reactor programme, since it is only through such a thermal programme that the plutonium needed to build up the fast reactor programme can be produced.*

Figure 9.5 *Some of the stocks of depleted uranium in store in the UK*

Once a fast reactor programme has reached its required total capacity, however, it can be effectively self-sustaining.

If Britain were to construct a Commercial Demonstration Fast Reactor (CDFR) before the 1990s in order to assess its commercial viability, she could (if all went well) begin to plan the introduction of full-scale commercial stations by the turn of the century, and could have up to 20,000 MW of fast reactor generating capacity in operation by the end of the first quarter of the 21st century. Taken along with her planned thermal capacity, this would allow Britain to produce between 50 and 60 per cent of her electricity from nuclear power – thus allowing large amounts of fossil fuels to be diverted to other uses and, by the middle of the century, she could (if she wished) produce virtually *all* her electricity from nuclear power by continuing to build up such a programme.

France, which will soon have its first CDFR in operation, has already embarked on an ambitious programme of this type and expects that, by 1990, up to 50 per cent of her electricity needs will be met from nuclear power. Because of the enormous electricity requirements of the USA, it seems unlikely that she will ever reach this level of commitment to nuclear power, but Britain, being in a less fortunate position in terms of natural resources, may have to move in this direction if she wishes to regain some of her earlier glory as a leading industrial nation.

Expansion of nuclear power, which is currently based solely on thermal reactors, will ultimately come to a halt unless the major industrialized countries act quickly. They can either begin to exploit the more expensive sources of uranium, which would inevitably reduce the economic attractiveness of nuclear electricity, or they can follow the lead of France, and build for the future by developing a role for fast reactors in conjunction with thermal reactors. A mixed programme of thermal and fast reactors of the type outlined above would guarantee supplies of electricity far into the next millenium – or until such time as nuclear fusion or solar energy has a proven technology which can take over from nuclear fission. Unless we embark on such a programme without delay, however, the option will disappear, since we will not be able to build up a self-sustaining fast reactor programme of the required size before the fuel for thermal reactors runs out. Thus, we should be planning for the future *now*; tomorrow will be too late.

PART FOUR

Social and Environmental Considerations

It has long been recognized that ionizing radiation is dangerous and potentially lethal. On this point, and on few others, the pro-nuclear and anti-nuclear lobbies are in agreement. However, the interpretation of the word 'dangerous' seems to depend on which standpoint one adopts. On the one hand, those who oppose nuclear power argue that *any* increase in radiation exposure is dangerous, however small it might be, and should therefore be avoided at all costs. On the other hand, those who support nuclear power argue that the effect of the nuclear power industry on the background radiation level is so small as to be insignificant, and, in any case, is always likely to be masked by natural and regional variations. A very large part of the debate that has been raging ever since the first nuclear power station was planned has been concerned with this basic difference in interpretation.

In Parts II and III of our book, we have tried to give objective answers to the questions: 'do we need nuclear power?' and 'what sort of nuclear power programme would best meet our future requirements?'. In this final part, we will examine the more subjective aspects of nuclear power, and will try to answer what is perhaps *the* most important question, namely 'is nuclear power socially and environmentally acceptable?'.

As we said in the Introduction, many of the books that have been written about nuclear power only give one side of the story. By ignoring the balancing arguments that can almost always be made by the opposition when examining any particular issue, the 'evidence' can generally be made to appear overwhelming – irrespective of which side of the case is being presented. Over the past ten years or so, the public have therefore been swayed backwards and forwards between the two sets of protagonists, who have simply presented their own cases, citing only the 'evidence' which most strongly supports their particular point of view. Governments, the power industries and the nuclear authorities must take much of the blame for this unsatisfactory state of affairs, since they have, until recently, been extremely reluctant to open their books to public eyes, and have assumed that the whole area of nuclear power is 'too technical' to be placed before the general public. This situation is now slowly changing, however, and one significant result of this change in attitude by the authorities has been a major shift in the emphasis of the anti-nuclear arguments. Most informed anti-nuclear bodies now consider that the question of nuclear proliferation (with which we will deal in Chapter 12) is perhaps *the* most important unresolved issue. For the specialists, this may or may not be so, but for the general public, a continually shifting battleground is no more satisfactory than the earlier situation, and no more likely to enable them to make up their own minds on a crucial issue that will undoubtedly have considerable impact on all our lives – however it is eventually resolved.

In Part IV, we will examine the various social and environmental aspects of nuclear power by looking at *both* sides of the argument. We will do this by taking each area in turn, and presenting, in summarized form, the cases for and against nuclear power. Where appropriate, we will take a detailed look at the evidence and offer our own interpretation (eg in the case of the Three Mile Island incident). By the end of these last three chapters, we hope that you, the reader, will be in a better position to decide for yourselves whether *you* consider the continuation of national nuclear power programmes to be socially and environmentally acceptable.

CHAPTER TEN

The Hazards in the Nuclear Power Industry

One of the major reasons for public concern about nuclear matters in general, and nuclear power in particular, is that ionizing radiation cannot be detected by the human senses. This, in effect, places nuclear power in a different psychological league from, say, coal-mining or the chemical industry, and, as a result, people find it very difficult to be objective about the safety (or otherwise) of nuclear power.

The general public are also very concerned about the link between radiation exposure and cancer. One of the sad things about the nuclear debate in the past has been the sensationalism of the media, who have frequently been seen to adopt similar tactics to the anti-nuclear lobby by reporting only that material which shows the nuclear power industry up in a bad light.

We will begin this present chapter with some thoughts and comments about radiation levels and the link with cancer. Since these sensitive matters are common to all sections of Part IV, we hope that we will be able to clear up several misconceptions right at the beginning.

The group of people who are undoubtedly most at risk from the nuclear power industry are the members of its workforce. The nuclear power industry begins in the uranium mines, and ends, at the moment, in the reprocessing plants. We will therefore take a look at the potential dangers to the various sections of the workforce, and at the measures taken to safeguard health. All too often, the dangers of working in the nuclear power industry are considered in isolation, so we will take this opportunity to compare its safety record with those of other industries.

Radiation levels and the link with cancer

At the beginning of Chapter 4, we said that many of the early scientific workers in the fields of radioactivity, X-rays, etc, subsequently died of various forms of cancer that were attributed to the radiation to which they had been exposed. Madame Curie, who spent a major part of her life working with radium, is perhaps the best-known victim.

Cancer is a very unpleasant disease which carries with it an inevitability that is not associated with many other diseases. There is no doubt that radiation can and does cause cancer. What is not quite so clear is whether there is a threshold below which there are no adverse effects (as is often the case with toxic chemicals that produce acute effects). At present, it is generally assumed that there is *no* safe threshold below which there is no effect, but, at the very lowest dose levels, it is assumed that the additional risk is completely masked by the 'normal incidence rate', and is therefore statistically undetectable.

Since 1910, as evidence increasingly linked cancer to exposure to ionizing radiation, the International Commission on Radiological Protection (ICRP) and its forerunners have gradually reduced the permissible exposure levels. At present, the radiation safety levels are set at five rems per year for workers in the nuclear power industry and 0.5 rems per year for members of the general public (see Chapter 4 for the definition of the rem). The radiation exposure from natural sources is roughly 100 millirems per year (depending on where you live), and to this we must add roughly 50 millirems per year from medical and other man-made sources. (A more detailed breakdown of these figures is given in Chapter 11.)

By assuming that even the smallest levels of radiation can have some effect, it is possible to put the cancer risks produced by exposure to radiation into a proper perspective. As a result of extensive studies, it has been estimated that roughly 10,000 man-rems are required to produce one cancer death. On the basis of this linear hypothesis, we can say that there would be the same effect if 10,000 people each received one rem of radiation, or 1000 people each received ten rems of radiation, or 100 people received 100 rems, and so on. In other words, any combination of numbers of people and dose level that produces the figure of 10,000 man-rems is, according to current medical opinion, likely to cause one cancer death.

Over the past few decades, there has been a noticeable increase in the proportion of deaths from cancer. At the same time, there has been no significant increase in the background radiation levels. Is the above linear hypothesis therefore wrong? We do not really think so, since there are at least two obvious causes for this increase. Firstly, the use of antibiotics has greatly reduced the number of deaths from other causes, and so more people are now dying of cancer rather than, say, pneumonia. Secondly, deaths from lung cancer brought about by cigarette smoking have reached almost epidemic proportions in Western society.

To put cigarette smoking into its proper perspective, the expected reduction in life span from smoking is compared with other known life-reducing factors in the table in Figure 10.1. We see from this list that the average man or woman who smokes 20 cigarettes a day for 50 years will reduce his or her life by roughly 3½ years, or five minutes for every cigarette smoked. Thus, the average reduction in life span that would result from *all* our electricity being produced from nuclear power is roughly equal to that produced by smoking seven cigarettes!

If the linear hypothesis is correct, then we can assume that the background radiation dose will be responsible for a proportion of the total cancer deaths. Taking the population of the USA to be roughly 220 million and the average background radiation level to be roughly 150 millirems, the annual rate of 'natural radiation-induced cancer deaths' comes out at roughly 3300. A similar calculation for Britain produces a figure of roughly 840 natural radiation-induced cancer deaths per year.

Factor reducing life span	Statistical average reduction in life span
Being 25% overweight	3.5 years
Smoking 20 cigarettes a day for 50 years	3.5 years
Being born male rather than female	3 years
Average natural lifetime radiation dose	3 days
All our electricity being generated from nuclear power	36 minutes
Smoking *one* cigarette	5 minutes

Source: Wyatt, A (1978) *The Nuclear Challenge — Understanding the Debate.* Toronto: Book Press Ltd

Figure 10.1 *Reduction in life span from different causes*

The total annual cancer death rate in each one million of population stands at about 1600; this figure includes all forms of cancer, and all the various causes of cancer. For the USA, this means that roughly 350,000 cancer deaths occur each year; for the UK, the corresponding figure is nearly 90,000.

Now that we have these figures available, it is possible to demonstrate the dangerously one-sided nature of the anti-nuclear arguments and the media treatment of 'nuclear radiation':

1. Quite correctly, the anti-nuclear lobby suggests that there is no such thing as a *safe* dose of radiation, and that any dose, however small, is potentially harmful. In arguing that we should therefore be very concerned about radiation from nuclear power generation, they cite the situation in Japan, where, it is claimed, over 100 people still die every year from the effects of the atom bombs that were dropped in 1945. Let us take a closer look at this claim.

   Of the 285,000 survivors at Hiroshima and Nagasaki, 82,000 have been carefully monitored for more than 30 years. Of those in the study, 4000 have since died from cancer. The number of cancer deaths that would have been expected in an *un-irradiated* population of the same size over the same period is roughly 3800, however. Thus, it appears that only about 200 (or one in 20) of these cancer deaths can be directly ascribed to the radiation from the bombs, in spite of the large doses of radiation received.

2. An American TV programme entitled *Paul Jacobs and the Nuclear Gang* (shown also in the UK) dealt with the final days in the life of a reporter who was dying from cancer. This was a particularly harrowing programme in which the reporter attributed his cancer to radiation he had received many years ago, while reporting on American nuclear weapons trials. The man subsequently died of lung cancer.

   What was *not* stated in that programme was the fact that Paul Jacobs was a chain smoker who had smoked 50-60 cigarettes a day for most of his life. Obviously, it is impossible to say how his cancer originated, but this extra piece of information puts his illness and subsequent death in a perspective that is decidedly different from that gained from watching the programme.

The fact is that there are *many* causes of cancer, and exposure to ionizing radiation is only one. There are a large number of chemical agents that are known to cause cancer; these are called *carcinogens* and most of them are *far* more dangerous than the radiation from nuclear power stations. Many hydrocarbons, for example, are highly carcinogenic and, unfortunately, several of these are quite common. Benzopyrene, for instance, is produced in the burning of coal, in cigarette smoke and in the exhaust fumes of motor vehicles. In our opinion, bodies who are concerned about the dangers of cancer would do far more good if they switched their attacks to the sources of these chemical carcinogens.

Are radiation exposure limits too high?

It is often pointed out that the radiation safety levels for workers in the nuclear power industry are ten times higher than those for the general public and over 30 times the dose that they receive from natural causes. In addition, the situation at Windscale (UK) is often used to demonstrate the unacceptably high nature of this limit. At Windscale, the total dose sustained by the workforce was 12,000 man-rems in 1976 and 8000 man-rems in 1979. It is claimed that this is unsatisfactory, since it will probably result in at least one cancer death per year; thus (it is argued) a large number of Windscale workers have already been condemned to premature death from cancer.

In attempting to answer the critics on this point, it must first be said that the dangerous nature of ionizing radiation is not in dispute. The nuclear power industry has always been aware of the dangers and has, at all times, taken the most stringent precautions to ensure that workers are put to minimum risk. It is also accepted that there is no such thing as a completely safe dose of radiation, but, for doses below a certain critical value, it appears that there are no detectable short-term or long-term harmful effects. The dangers at these low-dose levels appear to be negligible compared with the other risks that the workers are subjected to in the course of their normal lives. The radiation safety limits for workers in the industry are, of course, set by the ICRP. This is an independent body that is concerned with *all* sources of ionizing radiation, and was established in 1928 – long before nuclear fission was even discovered.

It should be stressed that the radiation safety limits are *maximum* limits. Very few workers are in fact ever exposed to anything like these doses. In 1977, for example, the CEGB (UK) had 5254 'classified' personnel who worked in areas where they were liable to be exposed to some radiation. Of these, over 85 per cent (4477) received whole-body doses of less than 0.5 rems, 14 per cent (734) between 0.5 and 1.5 rems, and less than one per cent (39) between 1.5 and five rems. *No* worker received more than the ICRP limit of five rems. A similar picture emerges from the other side of the Atlantic. For example, the Ontario Hydro average for all its nuclear power station workers is less than one third of the permitted maximum limit.

Any worker who is exposed to one rem of radiation runs a one in 10,000 risk of subsequently contracting cancer as a direct result of that exposure. This is roughly equivalent to the cancer risk produced by smoking three cigarettes a week.

In the past, considerable concern has been expressed regarding the risks to which uranium miners are exposed. Apart from normal mining accidents, the main hazard comes from the inhalation of the radioactive gas radon, which is formed during the radioactive decay of radium – an element that is present to a greater or lesser extent in all rocks and soils but particularly in uranium ore. The alpha particles that are released from radon inhaled over a long period can deliver to the lungs a sufficiently high dose of radiation to cause lung cancer.

A survey undertaken in the USA in the 1950s showed that there was a clear relationship between the excess frequency of lung cancer in uranium miners and the average radon concentration in the mines in which they had worked. The excess death rate from lung cancer in uranium miners was found to be roughly 15 deaths per 10,000 man years of mining. This figure, of course, relates to the mining conditions that prevailed many years ago – when ventilation techniques were crude and miners were an exploited minority. In recent years, conditions have greatly improved, and it has been estimated that, in terms of the number of deaths associated with the production of a given quantity of electricity, the mining of uranium is now only about one tenth as dangerous as the mining of coal. This is largely due to the fact that far fewer miners are needed to mine the uranium that is required to generate a given quantity of electricity than are needed to mine the coal required to generate the same amount.

What is often forgotten is the additional problem that the coal industry has with pneumoconiosis. In the USA, ex-coal miners are dying from this disease at the rate of almost 80 every week, and there are estimated to be nearly *half a million* coal miners with the disease. In the UK, the corresponding figure is around 30,000 miners and ex-miners.

To date, the use of nuclear electricity in the UK is thought to have saved the lives of roughly 50 coal miners. In Canada, it is estimated that several hundred miners will be saved from fatal accidents and roughly ten times that number from serious injury or disease between now and the end of the century if the proposed nuclear power programme is allowed to go ahead.

We can only hope that *all* mining operations become safer in the future, because both uranium and coal are going to be needed for a long time to come.

Is the safety record of the industry acceptable?

In the routine operation of the nuclear power industry, the workforce is protected by the ICRP limits, all members of the workforce being regularly monitored and health checks being mandatory. Practically all the radioactivity that is produced is concentrated in the actual fuel elements, and either ends up in the cooling ponds where spent fuel elements are stored (in the once-through fuel cycle) or as radioactive waste from a reprocessing plant (in the other fuel cycles). In neither case does it represent any hazard to the workforce unless there is some sort of accident that results in the release of radioactive materials into the working environment. Let us therefore try to assess the extent to which the workforce are at risk from such accidents, which could, of course, be very serious indeed should a massive leak occur.

When the safety record of the nuclear power industry is under scrutiny, the anti-nuclear lobby often talk of 'accidents', whereas the nuclear power industry prefers to use the word 'incidents' – a good example of the semantic relativism that so frequently fogs the nuclear issue. The main arguments that are made against the nuclear power industry in this respect can be summarized as follows:

1. The nuclear power industry claims that it has an excellent safety record. This, it is argued, is manifestly not the case; many nuclear accidents have in fact occurred, but the great majority of these are never revealed to the general public. For example, the US government's Nuclear Safety Information Center at Oak Ridge recently disclosed that they had investigated no less than 2000 'incidents' in 1979, of which no fewer than 32 might have resulted in a catastrophic reactor meltdown. The 1979 Three Mile Island incident is a good example. Here a series of accidents, mistakes and downright blunders led to a near disaster that put many plant workers at risk – not to mention the general public.
2. A major accident, resulting in the release into the atmosphere of large amounts of radioactive material, occurred at Windscale in 1957. In terms of the amount of radioactive material released, it is argued that this was far worse even than Harrisburg, resulting in all the milk produced in the surrounding countryside being declared unfit for human consumption. It is also claimed that the radiation released from Windscale even reached Alaska!

3. The nuclear power industry claims that no one has ever died as a result of the British civil nuclear power programme. It is argued that this again is manifestly not the case. Indeed, since the Windscale public inquiry, it is pointed out that there have been no less than three deaths of workers for which British Nuclear Fuels Ltd have paid compensation to the widows.
4. It is openly admitted by the nuclear power industry that it is not possible to build a completely safe reactor. Thus (it is claimed) a serious accident that puts the workforce (and the general public) at severe risk is inevitable sooner or later.
5. It is argued that workers in the nuclear power industry are put at special risk because they are exposed to plutonium – the most dangerous substance known to man (so it is claimed).

Since much of the criticism raised in the above paragraphs is concerned with the effect of the nuclear power industry on the general public as well as on the workforce, we will leave a detailed discussion of the Three Mile Island and Windscale incidents to Chapter 11. Here, we will concentrate on the points directly relating to the safety of the workforce.

Despite the above claims to the contrary, we believe that the nuclear power industry has an excellent safety record that can be favourably compared with that of any other industry. As in any other industry, accidents do occur, but, for some reason, the nuclear power industry is always selected for special mention. This is, perhaps, because of the psychological effect of radiation – the fact that it cannot be seen, heard or smelt. Whatever the reason, it seems to us that much more effort could be profitably directed towards increasing the safety records of many other industries. The table in Figure 10.2 illustrates our point. It is taken from the 1975 Health and Safety Statistics for the UK. Similar comparisons can be made for other nations with nuclear power programmes with very similar results.

The basic protection afforded to the workforce in the nuclear power industry includes the following elements:

(a) All nuclear plants are designed to reduce the potential radiation hazard to its lowest practicable level by the provision of safeguards which include multiple containment and radiation shielding (see Chapters 2 and 11).
(b) Safety rules which ensure that the maximum permissible radiation dose limits recommended by the ICRP are not exceeded are rigidly enforced.
(c) Sites are classified into different radiation zones according to the level of radiation that exists. Access to the various zones is closely controlled, and is restricted to certain classes of personnel. Also, the time for which workers are allowed to work in radiation zones is limited, so that no worker receives more than the permitted dose.
(d) Contaminated zones are always clearly defined, and access to these is strictly controlled. Special procedures (protective clothing and,

Industry	Accidents per 100,000 employees	
	Fatal	*Total*
Shipping (merchant)	86.0	1800
Coal-mining	23.4	19,753
Coal and petroleum products	22.4	6570
Construction	19.4	3660
Railways	18.7	2920
Ship-building and marine engineering	14.0	6150
Agriculture	11.7	1800
Metal manufacture	10.0	6350
Bricks, pottery, glass and cement	7.2	4750
Chemicals	6.7	3640
Timber and furniture	5.5	3280
Mechanical engineering	3.4	4110
Paper printing and publishing	2.9	2270
Electrical engineering	1.8	2320
Food, drink and tobacco	1.7	4370
Textiles	1.7	2750
Nuclear power generation	0.0	2897*

* All from non-nuclear causes

Figure 10.2 *Accident statistics for different industries – UK (1975)*

on occasions, breathing apparatus) may also be specified for certain tasks.

(e) All employees who are occupationally exposed to radiation are 'classified'. All such classified personnel wear film badges, which are regularly checked in order to monitor radiation dosage, and also undergo regular medical and blood checks. No person is allowed to leave a contaminated zone until he has been through a personal monitoring procedure. Comprehensive records are kept of the doses received by all workers.

(f) Routine radiological surveys are carried out as an extra safeguard against exposure to radiation.

With regard to the 1979 Harrisburg accident, which is now frequently cited by the anti-nuclear lobby, the leakage of radioactive materials from the core of the reactor was almost entirely contained within the reactor building, the hazard to the general public being practically zero. Also, no worker was killed or even injured by the incident. The maximum exposure received by an individual was believed to be 37 millirems (about as much exposure as he would have received from natural background radiation in four months). The reactor actually shut down automatically, as it was supposed to, and there was little chance of a 'catastrophic meltdown' occurring.

In the case of the 1957 Windscale incident, the committee that was set up to investigate the accident subsequently reported that 'it is in the highest degree unlikely that any harm has been done to the health of anybody, whether a worker in the Windscale Plant or a member of the general public'. The report also stated that the incident had produced no radiation or inhalation hazard in the surrounding area. The much-publicized withdrawal from the market of locally produced milk was carried out mainly as a precautionary measure to allay public fears. Finally, with regard to the claims that radioactive materials released by the accident 'reached Alaska', the instruments available at the time were simply not sensitive enough to make any meaningful measurements, so that such statements should be dismissed as malicious nonsense.

As we have already stated, workers in the nuclear power industry have a slightly higher than normal risk of contracting cancer – a one in 10,000 chance for every rem of radiation to which they are exposed. There is, however, a possibility that the linear hypothesis on which this figure is based is more likely to overestimate rather than underestimate the effect at low dosages. It is well established that there are mechanisms in the body for repairing radiation damage. Chromosomes broken by radiation have been seen under a microscope to reunite, and there is also evidence that the latent period before cancers develop (roughly 15 years for large exposures) increases with decreased exposure.

Despite these reservations, compensation was in fact recently paid to the widows of three Windscale workers who died of cancer and, in two cases, it was accepted that the disease was on the balance of probabilities induced by radiation received at work, although it is difficult to prove this point one way or the other.

Finally, we come to the question of plutonium – 'the most dangerous substance known to man'. While admitting that plutonium is highly toxic, we would say that it is in fact considerably less so than several other substances which are regularly used in a wide variety of industries without any particular fuss or protest being made. Lead arsenate, for instance, is over *ten times* as toxic as plutonium when ingested, and a lethal dose generally produces death in hours or days rather than years or tens of years as in the case of ingested plutonium. With regard to the dangers associated with plutonium, the following points might be borne in mind:

- It is *not* true that plutonium is highly radioactive. Its principal isotope, plutonium 239, is relatively stable and, as a consequence, its half-life is very long and its radioactivity per unit mass very low (see Chapter 4).
- It is *not* true that, in all circumstances, very small amounts of plutonium are lethal. Insoluble particles of plutonium or plutonium oxide are certainly hazardous when inhaled even in small quantities, but considerably larger amounts could be eaten without appreciable harm.
- It is *not* true that plutonium is only safe when protected by large

amounts of shielding. It could safely be sat upon by a person wearing no greater protection than a 'stout pair of jeans'. Plutonium is an alpha emitter, and, as we saw in Chapter 4, alpha particles are stopped by thin layers of plastic or paper.

– It is *not* true that an escape of plutonium would cause a disaster of unprecedented proportions. The damage done by the breaking open of a road or rail tanker of chlorine (for example) would be a great deal more damaging than the breaking open of a flask containing plutonium.

No industry, including the nuclear power industry, can expect to operate without *any* accidents, but some of the more irresponsible elements in the anti-nuclear lobby have been known to say that, since nuclear power cannot be made 100 per cent safe, it is potentially so dangerous that all nuclear power installations should be closed.

Nuclear power *is* potentially dangerous, like many other industries. Its safety record is, however, second to none. Admittedly, some expensive mistakes have been made, both in the USA and in Britain, but none of those which have occurred in the civil programmes have so far resulted in any real danger to the workforce or to the general public. Indeed, no worker in the British nuclear power industry has yet died as the direct result of a nuclear accident – a record that many other industries might envy.

CHAPTER ELEVEN

The Hazards to the General Public

In this, the penultimate chapter of our book, we will take a detailed look at the claims and counterclaims that are frequently made regarding the potential hazards that the various aspects of the nuclear power industry represent to the general public. We will look in turn at the hazards associated with the actual operation of nuclear power stations, the transportation of nuclear fuel and radioactive materials, the reprocessing of spent fuel and the disposal of radioactive waste. Finally, we will examine the question of whether the nuclear power industry represents a threat to the genetic future of our species.

In many instances, we will find that hard experimental evidence is not available because it does not exist, although in some cases, such as the Three Mile Island 'incident', enough evidence is available to allow an objective assessment to be made. Many of the fears concerning the safety (or otherwise) of the nuclear power industry are, we believe, based either on ignorance or on a genuine fear of what *might* be rather than on what has been shown to be the case. Because it is certainly true that some sectors of the nuclear power industry *could* represent real hazards to the general public, it is important to try and assess the nature and extent of these dangers. To do this, it is necessary to resort to a technique known as *risk analysis* – a mathematical method whereby the probability of a particular kind of accident occurring can be estimated. In order that the remainder of the chapter can be placed in its proper perspective, we will therefore begin with a brief, non-technical discussion of the nature of 'risk' and its relationship to safety.

How safe is safe?

No human activity can be guaranteed to have zero risk associated with it. For instance, it is now generally accepted that the smoking of cigarettes is linked with the incidence of lung cancer, and that there is a risk to one's life every time one crosses a road, travels by car, train or plane, swims or even climbs the stairs to go to bed. Most people are, however, prepared to accept the risks associated with activities like these because they are all individually perceived risks with an element of choice attached. The risks are also a part of our accepted way of life and, with the possible exception of cigarette smoking, a decision to avoid them at all costs would be a choice that very few people would be prepared (or able) to take.

Hazards such as floods, famine, plague, earthquakes or meteorite impact that are outside human control are also acceptable, if only because they are difficult or impossible to avoid.

It is when society as a whole is confronted with the associated risks of any technological development that it neither fully understands nor has effective control over that doubts and conflicts arise. Consider what would happen if, for example, a present-day government recommended the widespread use of a transportation system that would provide highly convenient personal transport but would kill at least one in 8000 of the population every year. We are sure that pressure groups would mount a relentless campaign opposing such an innovation because, it would be argued, any technological development that was likely to kill upwards of 125 people per year out of every million of the population could *never* be justified, whatever the benefit to society. We are, of course, talking about the motorcar, which is (we would argue) only accepted by society because it has been around for a long time, so that we are used to it, and would not be prepared to suffer the inconvenience of doing without it.

To say that a certain activity or industry is 'safe' is simply to imply that the risk associated with it is low enough to be considered either negligible or socially acceptable; 'safe' can *never* mean 'zero risk'. In some cases, it is possible to use accident statistics to obtain a quantitative measure of risk, but the interpretation of the results of such an exercise must always be treated with caution. By way of example, let us look at the figures for accidental death in the home. In the UK, accidents in the home kill between 5000 and 6000 people each year. In a population of 56 million, this represents an average annual risk of about one in 10,000. What is not clear from these figures is that roughly 80 per cent of the accidental deaths are among people aged over 65, so that the risk to this particular age group is five times as great as for the population as a whole.

A similar type of picture emerges if we consider deaths from road accidents. Although the overall risk is calculated at roughly one in 8000,* it appears that twice as many men as women die in motor accidents. There are obvious explanations for these differences, which are clearly important for the individual but perhaps not so important for society as a whole. In fact, human nature tends to make us question the relevance of such statistics: after all, how many people over the age of 50 believe that they are less healthy than others they know, and how many drivers accept that their driving ability is only 'average'?

'How safe is safe?' is therefore a question that will produce different responses from different individuals; it depends on whether the activity under consideration is seen to affect the individual directly or indirectly;

* This is the figure for the UK; the corresponding figure for the USA is almost twice as high.

and whether or not he or she has an intimate understanding of it. This is most clearly demonstrated in the case of the nuclear power industry. Most people working in the industry consider its safety record to be second to none, and cannot understand why members of the general public and, in particular, members of the anti-nuclear lobby, are so concerned. Perhaps the reason is that the adverse publicity given to the industry influences the people involved rather less than it does those who have no specialized knowledge. Or perhaps the reason lies in the fact that although no one has yet been killed in a nuclear accident in a commercial nuclear power station, if there *were* to be a 'nuclear disaster', then thousands of people might ultimately die.

Because of the nuclear power industry's excellent safety record, the risks involved can only be estimated by evaluating the probabilities and likely consequences of all conceivable accidents. This approach, known as *risk analysis*, is an attempt to put the whole subject of safety on a more scientific footing. The technique is briefly described below, using a nuclear installation as an example of its use.

The risk analysis approach

The techniques of estimating risk by probabilistic analysis was originally developed during the 1960s by the UKAEA, who have since used it to guide their choice of reactor safety research programmes. In 1975, it was also used in the USA to assess the accident risks in US commercial nuclear power plants – a study directed by Professor Rasmussen at Massachusetts Institute of Technology (WASH 1400: see Bibliography).

The first step in the analysis of the risks associated with a nuclear installation is to identify every conceivable initial failure within the plant. Next, the technique is used to examine all subsequent courses of events that could result from this initial failure and identify any that could eventually lead to an uncontrolled release of radioactivity into the environment. This is done by a logical process known as 'fault tree methodology'. Fault trees can be used to identify thousands of potential accident paths.

The next stage in the process is to analyse each fault tree in order to determine the likelihood of failure of the various systems identified therein. Beginning with the initial failure, it is then possible to use engineering and mathematical logic to determine the ways in which each system in the tree could fail. In each case, a probability is allocated to the mode of failure. Using data on the failure of components (pumps, pipes, valves, etc), the likelihood of operator errors and the likelihood of maintenance errors, it is then possible to predict a total probability for a whole sequence of undesired events (an accident) which results in the release of radioactivity into the environment, even though such an accident has never occurred in practice.

The final step is to estimate the probable consequences of the releases of radioactivity that would result from the various accidents. Since the main hazard arises from airborne releases, prevailing weather conditions, population densities and land usage are all taken into account. Using weather and population statistics, the number of injuries and fatalities is estimated in relation to the overall probability of each type of accident.

Fault trees are also widely used in the aeronautical industry as a tool for safety and reliability analysis, and official pressure is now being brought to bear on other industries (such as the chemical industry) to use the technique more widely in order to improve safety.

With this kind of analysis available, it is possible to identify major sources of hazard and redesign or protect potentially dangerous systems in order to reduce the overall risk and hence increase overall safety. The number of accidental deaths from any activity or within any industry could thereby be reduced, but the cost of achieving even a small improvement might be prohibitively high. Safety, in other words, can only be bought at a price. The more money that is spent on safety, the less we have to spend on other things. It is therefore important that, whatever money is made available for safety, it should be spent in such a way that it produces the maximum beneficial effect. How this is achieved is really a matter of assessing social and economic priorities. For example, it *would* probably be possible to make travelling by car as safe as the nuclear power industry, but it is almost certain that the measures that would have to be introduced would be totally unacceptable on the grounds of increased inconvenience and greatly increased cost.

The potential hazards to the general public represented by the nuclear power industry

As stated in the introduction to this chapter, we will now take each sector of the nuclear power industry in turn and review the arguments that are frequently put forward by the anti- and pro-nuclear lobbies. Because of the nature of most of the attacks on the industry, the pro-nuclear arguments generally take the form of a point-by-point reply to the various criticisms that are made, and we will adopt the same approach throughout the remainder of this chapter. It is fair to say, however, that while both sides acknowledge that potential hazards exist, they disagree as to whether or not the safety record of the industry suggests that the general public can live at ease with nuclear power.

The hazards associated with the actual operation of nuclear power stations

The anti-nuclear case

When discussing the safety (or otherwise) of nuclear power stations, the anti-nuclear lobby tends to focus its attention on the number of nuclear accidents (or 'incidents') that have occurred over the years, and concludes that the industry is manifestly not safe and is therefore socially unacceptable. Their case can be summarized as follows.

The supporters of nuclear power frequently try to alleviate public fear of serious accidents at nuclear power stations by claiming that the probability of such accidents is extremely low and that, since very few deaths have occurred as an obvious result of nuclear accidents, there have in fact been very few such accidents. The anti-nuclear lobby suggests that both these claims are misleading, and that the second is downright dishonest.

Taking the low probability claim first, they argue that even the nuclear power industry admits that there is a chance of roughly one in a million of a major accident occurring for every reactor year of operation. Despite the fact that, on paper, this is a fairly low risk, it is still unacceptably high in view of the terrible consequences that such an accident could have. If a power station suffered a complete meltdown, or one or more of its reactors blew up, releasing its radioactive material into the atmosphere, tens or even hundreds of thousands of people could conceivably die as a result. The exact number would depend on the nature of the accident, the position of the station with respect to population centres, and the direction of the prevailing wind at the time. Such an accident is so horrifying that even a one in a million chance is far too high. For nuclear power to be socially acceptable, they maintain that the chances of such an accident would have to be *zero*.

On the claim that very few nuclear accidents have in fact occurred, the nuclear power industry (particularly in Britain) is accused of hiding behind a tight wall of secrecy regarding its safety record, categorically refusing to disclose any of its safety studies to the public gaze. The comparison is made with the USA, where the situation is such that the nuclear industry is compelled by law to publish full information about any accident that takes place in a nuclear complex. The picture that emerges from these reports is said to be horrifying, revealing a whole series of accidents that have claimed the lives of technicians and resulted in large-scale contamination of reactor sites and their surrounding areas. In a rare lapse into statistics, the anti-nuclear lobby quotes accident figures for the USA. In 1969, for example, when there were 91 reactors and 314 research piles in the USA, no less than 120 suffered accidents of one form or another. Again, in 1979, the Nuclear Safety Information Center set up by the US Government at Oak Ridge is quoted as investigating no fewer than 2000 'incidents', of which no less than 32 might have resulted in a catastrophic core meltdown. That such a meltdown has not so far occurred is considered to be more a matter of good luck than good management. The American nuclear power industry is accused of poor design, faulty construction, incorrect operating procedures and even downright incompetence and carelessness.

Several specific accidents are often quoted in support of this case – for example, the incidents at the Enrico Fermi fast reactor, Michigan, in 1966, at Browns Ferry, Alabama, in 1975, and, most recently, at Three Mile Island, Harrisburg and Crystal River, Florida, in 1979. Long-term problems such as the discovery of cracks in some French PWRs and in some of the British Magnox stations are also highlighted as examples of the dangers inherent in any nuclear power programme. In particular, it is claimed that, at Harrisburg, a combination of accidents, incorrect operating procedures and human error almost led to the complete meltdown of the reactor and to a major nuclear disaster that could have killed hundreds of thousands of people.

Another major cause for concern about the operation of nuclear power stations is the question of what will happen to them at the end of their operational lives. It is assumed that it is not possible to dismantle the stations because of the highly active nature of their reactor cores. Are they simply to be left standing, representing a permanent danger to the general public and potential targets for terrorist attack? This would require them to be guarded for ever. Or are they to be buried under artificial hills, as is sometimes proposed? Surely, it is argued, this would constitute an unacceptable danger to the environment owing to the probable leakage of radioactive materials into the ground and the resulting contamination of the water supply. No satisfactory solution to the disposal of nuclear power stations has yet been proposed (so they maintain) and, until such a solution is produced, we should not allow any more to be built.

The pro-nuclear reply

In replying to these criticisms, the nuclear power industry concentrates on trying to explain how safety is built into all nuclear power stations. Their task is not easy because, in many cases, a full explanation would require a detailed technical description that would be lost on the vast majority of people. The general form of reply can, however, be outlined as follows.

All nuclear power stations are designed in such a way that they are as safe as it is humanly possible to be, and are built to standards that are unequalled in practically any other industry. All stations have a whole range of built-in safety features, designed to back one another up should failure of an individual system occur. These include:

- the form of the fuel itself, which prevents an escape of all but a tiny fraction of the fission products even if rupture of a fuel can occurs;
- the cladding, which is designed to ensure that no fuel or fission products escape from a fuel can except in the most serious accidents;
- the coolant circuit, which ensures that the reactor cannot overheat even if partial failure of the system takes place;
- the various containment structures such as the reactor vessel, biological shield and the reactor building itself, all of which are designed to back one another up (the principle of *multiple containment*);
- various emergency shut-down systems;
- various back-up safety systems.

For a major accident to occur, the safety systems of the station would have to break down *at every level*. In other words there would have to be:

- failure of the multiple instrumentation systems that are designed to detect and control faults as soon as they occur;

- failure to prevent the fault from developing to a dangerous level by taking appropriate remedial action;
- failure of the fuel elements to retain the fission products;
- failure of the various containment systems that surround the core of the reactor;
- failure of the various secondary containment systems.

The likelihood of this string of failures occurring in a single plant at one and the same time is very, very small, and has never happened in over 1500 reactor years of operation. Before a licence is granted for a nuclear power station in Britain, for example, the Nuclear Installations Inspectorate must be satisfied that the chances of a major fault occurring and not being contained by the various safety systems is less than one in a million for every year of reactor operation. By using the technique of risk analysis, it has been estimated that the probability of an accident causing more than ten premature deaths is less than one in two million per reactor-year, and that the probability of an accident causing more than 100 premature deaths is less than one in ten million per reactor-year. More serious accidents are even more unlikely. To date, no serious accident of this type has ever taken place anywhere in the world. There is, incidentally, no chance of a reactor exploding; the worst thing that could conceivably happen is complete core meltdown followed by rupture of the containment systems and release of radioactive material into the atmosphere. The accidents that *have* occurred have demonstrated the wisdom of the safety approach to reactor design, in that their after-effects have invariably been very small and, in most cases, insignificant. The Three Mile Island accident, for instance, about which we will have more to say later, released far less radioactivity into the atmosphere than did the ash dispersal from the Mount St Helen's volcano.

It is always difficult to envisage a risk of one in a million; the non-specialist does not live in a probabilistic world, and so such numbers tend to have little meaning. To put the figure into perspective, it is interesting to note that it has been estimated that a person undergoes a one in a million chance of dying by

- travelling roughly 50 miles by motorcar;
- travelling roughly 240 miles by air;
- working for one to two weeks in a factory;
- smoking 1½ cigarettes;
- rock climbing for 90 seconds.

These figures were first presented in the Windscale Inquiry Report of 1978. It is also interesting to compare the chances of dying as a result of a nuclear accident with the chances of dying from other causes. Figure 11.1 shows these estimates in tabular form for the UK. Similar sets of data exist for most other industrialized countries, and show the same basic trend.

Finally, the extremely low probability of dying from a nuclear accident can be demonstrated rather dramatically by the following example. Suppose that a person who lives near to a nuclear plant decided to reduce

Cause of death	Estimated risk per person per year
All causes	1 in 85
Natural causes	1 in 88
All accidents	1 in 3000
Road accidents	1 in 8000
Fall	1 in 9000
Fire	1 in 60,000
Drowning	1 in 80,000
Air accident	1 in 280,000
Rail accident	1 in 320,000
Lightning	1 in 10 million
Earthquake or flood	1 in 20 million
Nuclear accident	less than 1 in 100 million

Figure 11.1 *Annual death risks in the UK – by major cause*

his chances of early death by moving further away from the plant. If, as a result, he had to travel an extra distance of more than 270 metres to and from work, he would in fact *increase* his chances of premature death.

The anti-nuclear lobby frequently tries to alarm the general public by citing a whole list of incidents at nuclear installations. Although the great majority of such incidents have in fact occurred at research establishments rather than at power stations, the accident that took place at Three Mile Island on 28 March 1979 has provided them with fresh ammunition for a renewed attack on the integrity of the commercial nuclear programme (see Figures 11.2(a) and (b)). They claim that it was only by good luck that accidents such as that which took place at Harrisburg did not result in major disasters. According to the pro-nuclear lobby, this is a totally misleading argument, since complete failure of the type required to cause a major accident *has never occurred in any nuclear accident*. In all the cases cited above, for example (the Enrico Fermi, Browns Ferry and Harrisburg incidents), the various safety systems in operation at the plant almost completely contained the accident and prevented any serious contamination of the environment, loss of life or injury.

At Harrisburg, for instance, despite the claims of the anti-nuclear lobby, the 'one in a million' accident did *not* occur, and there was never a time during the whole incident when it was ever *likely* to occur. The reactor actually shut down correctly eight seconds after the initiating failure, and has remained shut down since that time. As a result, the rate of heat production had fallen to roughly 4 per cent of full power after one minute, and fell to 0.1 per cent after one month, by which time natural circulation had been fully established, ensuring a safe long-term state. There was no core meltdown, and no major breach of the containment system took place, so there was only a very small release of radioactive material into the environment. In the ten days after the accident, the total estimated additional exposure of the two million population in

Figure 11.2(a) *Three Mile Island nuclear power station at Harrisburg, Pennsylvania*

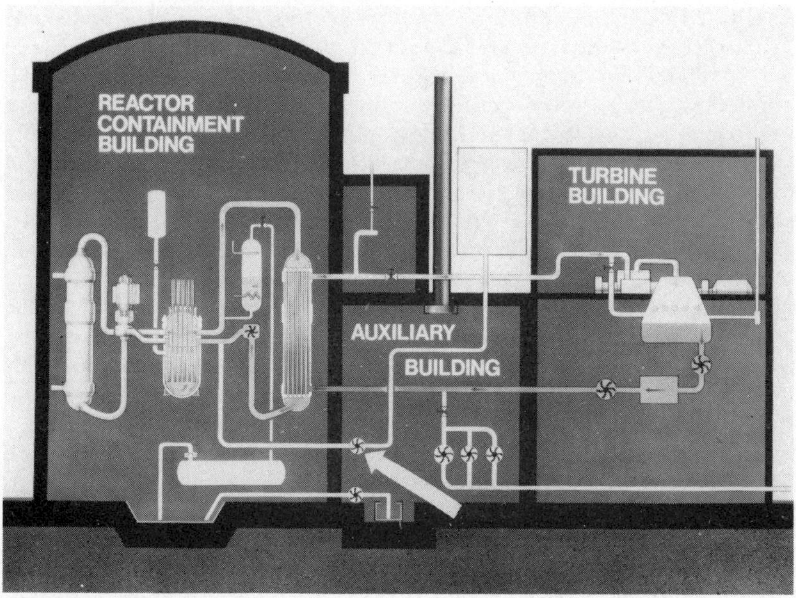

Figure 11.2(b) *A schematic diagram of the Harrisburg reactor showing the position of the pump in which the initial failure occurred (see arrow)*

the vicinity of the plant amounted to roughly 1/60 of the natural background radiation that they receive in the course of a year. As was stated in the previous chapter, no individual received a radiation dose in excess of 37 millirems (equivalent to four months' exposure from natural background radiation) and the average off-site exposure was only about 1.5 millirems. It has been estimated that the Harrisburg accident is eventually likely to result in (at most) one extra cancer death in the local population of two million, compared with an estimated natural incidence of roughly 325,000 fatal cancers. Indeed, it has been argued that the damage to the health of the local population produced by the actual incident was probably very much less than that induced by the stress that resulted from the hysterical media coverage and the uninformed scaremongering of the anti-nuclear lobby.

Although no one was killed or even injured at Three Mile Island, it was nevertheless a major financial catastrophe for the owners of the plant. Some 40 per cent of the fuel cladding is thought to have reacted with the steam (the reactors are of the PWR design), and, as a result, it is probable that virtually all the fuel pins have ruptured and large portions of the core have fragmented. The accident probably got out of hand because of a combination of weaknesses in detailed design safety analyses, failure of some components and operator error. If, for example, the operators had kept the emergency core cooling systems on through the early stages of the accident, it would have resulted in a relatively insignificant incident. A similar failure (in a pressurizer relief valve) in a similar type of plant at Crystal River, Florida, occurred later in the year. It appears that the lesson of Harrisburg had already been learned, because, on this occasion, the correct operating procedures were followed and, as a result, no damage resulted to the nuclear fuel in the core or to the plant.

It is certainly true that cracks have been detected in some PWR and Magnox systems. In the case of the Magnox reactors in the UK, the cracking is believed to have been present from the time of construction, and has only just been detected because of the availability of modern, sensitive equipment. The faults are being repaired by normal engineering processes, and the reactors will then be returned to normal service. The cracks in the PWR reactor vessels are estimated to have increased the chances of failure by something less than one per year in a sample of a million reactor vessels. The chances of a failure occurring before the end of the operational life of these reactors is therefore extremely remote.

In answer to the final point raised by the anti-nuclear lobby, the proponents of nuclear power state that it is in fact perfectly feasible to dismantle a nuclear power station *completely*, and to restore the site to its original 'green field' state. The decommissioning process is envisaged as having three separate stages. In the first, which is already a standard procedure, the spent fuel in the reactor will be removed, together with the coolant. The second stage will involve dismantling all the non-nuclear plant and instrumentation, thereby reducing the installation to the minimum practical scale without penetrating any part which has high levels of radioactivity. The final stage will involve removing the remaining plant, equipment and waste from the site. The site will then be released either for redevelopment or for general public use. This third stage is the one that is the most complex and costly, and will probably be delayed for several decades after stage 2, the actual period depending on the induced radioactivity in the reactor structure. Roughly 80 per cent of the material arising from the decommissioning will be inactive, and can therefore be disposed of in a conventional manner. Apart from the spent fuel, which will be stored prior to reprocessing, the remaining materials will have only low radioactivity (as a result of neutron bombardment during the operational life of the plant).

The hazards associated with the transportation of radioactive materials

The anti-nuclear case

Since fuel fabrication plants and spent fuel reprocessing plants are, effectively, centralized facilities, it is necessary to transport new fuel elements to reactors and spent fuel from reactors. Normal road and rail facilities are used for these operations, and this is a cause for concern among some anti-nuclear groups, particularly in the UK with its high population density. The main area for concern is the transportation of spent fuel elements.

Spent nuclear fuel is transported in huge metal flasks which can contain anything up to two tonnes of highly active substances. It is argued that the consequences of a serious accident involving one of these flasks, or an attack by terrorists using explosives or bazookas to blow one of them open, would be truly appalling. In the UK, there is one particular route that is used to transport these dangerous materials which actually runs through the suburbs of London. Opponents of nuclear power have estimated that if a major accident were to take place on this route, it could result in 8000 cases of cancer and the evacuation of an area ten miles in radius for a period of up to 70 years.

Doubts have also been expressed about the integrity of the flasks themselves, and about the stringency of the safety regulations which cover the transport of radioactive materials. It is suggested that the testing procedures for the flasks do not adequately simulate the conditions that might be experienced in an accident. In the UK, for example, the flasks are currently tested to withstand crashes at speeds up to 30 mph, yet they may apparently travel at up to 75 mph. With the possibility of a collision taking place at relative speeds of over 150 mph, this low speed test hardly seems adequate. Also, the flasks are designed to withstand fire for only 30 minutes, despite the fact that a major accident could easily cause a fire lasting for several hours. In support of their case, those concerned often quote a passage from the 6th report of the Royal Commission on Environmental Pollution, which says about the tests:

> We are surprised to learn that the tests are conducted only on models, and, since containers have to travel on freight trains which may be expected to travel at speeds up to twice that assumed in the tests, we were not really reassured.

There have in fact been a number of accidents involving the transportation of nuclear materials – collisions, derailments, etc – and again it is argued that these could so easily have ended in disaster. The recent introduction of ships to transport liquid plutonium nitrate from Dounreay in Scotland to Windscale has given the campaigns of anti-nuclear and environmental groups an added impetus.

The pro-nuclear reply

In replying to these various criticisms the supporters of nuclear power again point to the industry's safety record in connection with the transport of nuclear material, particularly when it is compared with the very real dangers that are associated with the transportation of chemicals. For example, they cite the accident that took place in Spain in 1977, when a road tanker containing a relatively small quantity of propylene gas crashed into a camp site and caused over 100 deaths and over 150 cases of serious burns. They also cite a similar accident that occurred in the USA in 1978 during the transfer of propane gas from a derailed tank car; here, an explosion caused 12 deaths and 50 serious injuries. Similarly, they cite an incident when an oil tanker, moored at a terminal at Bantry Bay in Ireland, suddenly exploded, producing 51 fatalities. It would in fact be possible to cite a whole list of similar accidents, but, instead of doing this, we will move straight on to the nuclear industry's replies to the specific points raised by its opponents.

A schematic diagram of the flask of the type used in the UK to transport nuclear fuel by road and rail is shown in Figure 11.3. The flasks weigh between 50 and 100 tonnes, and can carry up to two tonnes of irradiated fuel elements. They are designed in such a way that there is practically no chance of their being ruptured in any accident that is ever likely to occur. The equivalent American flask is designed to similar standards, and really only differs in its outward features. A photograph of an American spent fuel flask can be seen in Figure 11.5.

The proponents of nuclear power argue that the industry's safety record on transportation speaks for itself. In the UK, the transport of irradiated fuel by rail began in 1962. Between 1962 and 1980, over 4500 separate consignments, containing over 9000 tonnes of spent fuel, were carried a total of three million rail-miles. To date, there has been no accident to a flask that has resulted in any release of radioactive materials into the environment. In the very few minor accidents that have occurred, in no case was the flask involved subjected to even slight damage.

The British flasks are subjected to the following integrity tests:

(a) They are dropped on to a solid, non-yielding steel and concrete base from a height of roughly nine metres, in a number of different attitudes.

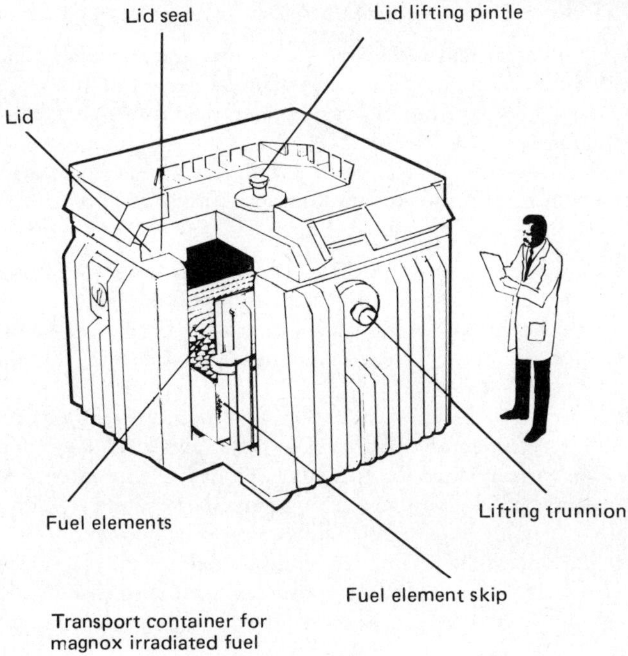

Figure 11.3 *Schematic diagram of a transport flask*

(b) They are then exposed to an intense petrochemical fire on all six faces (ie totally engulfed in flames) for 30 minutes, after which they are left to soak in unburnt fuel for three hours before any fire-fighting measures are applied.
(c) They are immersed in 15 metres of water for several hours.

In answer to critics who say that these tests do not represent the conditions to be expected in a serious accident, it is argued that the impact tests are in fact more severe than the conditions that any conceivable accident might produce. First, the container is dropped in those attitudes that are liable to produce greatest damage, and is, furthermore, dropped on to a structure that is much heavier and very much more rigid than anything with which the container is ever likely to collide. In a real-life accident, the flask would also receive some protection from the structure of the transporting vehicle. Second, it is extremely unlikely that an accident would ever produce an intense fire lasting more than 30 minutes. Such fires only occur in a very small proportion of real accidents (eg those involving aircraft or fuel tankers) and, in such cases, the inflammable materials are practically always dispersed by the initial impact so that the duration of the fire is limited.

Support is given to these arguments by tests which were carried out by Sandia Laboratories in the USA. The tests were at least as severe as those carried out in Britain, and involved the following elements:

(a) A flask on an articulated lorry was crashed into a massive concrete abutment at 60 mph, followed by a similar test at 80 mph.
(b) The flask, now on a railcar, was again crashed into a concrete abutment.
(c) The flask was placed on a road transporter on a level crossing and hit by a diesel locomotive travelling at 80 mph.
(d) Finally, the flask was engulfed in a petrol fire for 90 minutes.

Test (c) is shown taking place in Figure 11.4, and the end result is seen in Figure 11.5. As can be seen from the latter photograph, little damage appears to have been done to the flask, as was confirmed by the fact that no leakage of contents occurred either during or following the series of tests.

The claim by some anti-nuclear groups that nuclear fuel flasks could be burst open by terrorist attacks, resulting in widespread contamination, simply does not bear close scrutiny, it is argued. The contents of such flasks can never explode, and any attempt to scatter them by blowing up the container would require several tonnes of explosives so that the resulting explosion would itself cause massive damage to the surrounding area. Any terrorist group with the resources needed to produce such an explosion could, it is argued, choose much softer, more vulnerable targets than nuclear fuel flasks.

Figure 11.4 *A spent fuel flask being hit by a diesel locomotive travelling at 80 mph*

It is admitted that the use of smaller amounts of explosive might conceivably succeed in making a hole in a flask. This could lead to some escape of radioactive materials, and might also lead to a slow release of further amounts over a period of several hours. Emergency plans exist to cover all these eventualities, however, and it is extremely unlikely that such an attack would result in anything more than a very small escape of radioactive materials, and extremely unlikely that it would put the population of the area at risk.

A number of studies on the likely consequences of different types of accident involving the transportation of nuclear materials have been carried out in the USA. Although their conclusions differ in detail, they all seem to indicate that the chance of any fatalities resulting from accidents of this type is extremely low. If, for example, all America's electricity were produced from nuclear power, the number of deaths due to the release of radioactive materials in such accidents would probably be less than one per century.

The hazards associated with the reprocessing of nuclear fuel

The anti-nuclear case

As we explained in Chapter 3, nuclear fuel that has been irradiated in the reactors of nuclear power stations is ultimately transported in flasks to

Figure 11.5 *The spent fuel flask after the complete series of proving tests*

a reprocessing plant (provided, of course, that the once-through fuel cycle is not being employed). The contents of a spent fuel element depend on the detailed design of the reactor and the irradiation time, but a typical composition, on arrival at a reprocessing facility, might be:

	~	97% uranium
0.1	→	1% plutonium
2	→	3% fission products

On the subject of the chemical reprocessing itself, there seems to be little controversy. The chemical processes used in the reprocessing of nuclear fuels are used in a variety of other industrial situations, and the technology is well understood and, apparently, socially acceptable. As always, however, it is the presence of radioactivity, and how it is handled, that worries the anti-nuclear lobby. In the UK, the target for their attacks is BNFL's reprocessing facility at Windscale, and the individual criticisms raised against this facility are typical of the anti-nuclear stance on reprocessing.

Spent reactor fuel is dangerously radioactive, and, at first, generates large quantities of heat owing to the decay of short-lived fission products in the fuel. Before the actual reprocessing can take place, it is therefore necessary to store the spent fuel elements under water in order to allow them to 'cool off', ie to allow the most active fission products to decay. It is argued by the opponents of nuclear power that the metal casings of the spent fuel elements begin to corrode during this storage period, eventually releasing radioactive materials such as caesium 137 into the water. At Windscale, the pond waters are flushed into the Irish Sea, and, unless the storage period is short, this can (it is argued) lead to the discharge of dangerously large amounts of radioactivity into the sea. This is exactly what has been happening at Windscale (so the anti-nuclear lobby maintains) and, as a result, the Irish Sea is now 'the most radioactive sea in the world'. Evidence is quoted of fish caught locally containing nearly half the internationally agreed limit of radioactive materials, and of radiation pollution in a nearby town having reached 10 per cent of the limit. In the booklet 'Who needs Nuclear Power?' (SANE) a Professor Radford (described as 'the world's leading expert in the field') is quoted as saying that this level of pollution is 1000 times too high, and that the whole village should be evacuated *now*.

Fears are also expressed that the plutonium released from Windscale in its waste waters is steadily building up on the sea bed, and is gradually moving towards the land. Enough plutonium could (it is claimed) eventually build up to form a critical mass, and a small nuclear explosion could result. This would have a devastating effect on the area.

Several thousand tonnes of spent fuel from the British Magnox programme have already been processed at Windscale and the vast amounts of radioactive materials that have been discharged in the process are said to be a national disgrace. Even at their present levels, it is estimated that they will cause 20 additional cancers a year. Since the plant is shortly to be expanded so that fuel from the remainder of the British nuclear programme can be processed – as well as that from other countries such as Japan – it is feared that the problem is bound to become worse in the future.

The pro-nuclear reply

One of the most confusing aspects of the whole nuclear power issue for the non-technical observer is that both sides present statistical analyses, with the backing of 'world experts', that violently contradict each other. While it is true that the anti-nuclear lobby seldom tries to quantify its arguments, here is an example where such a conflict arises. In trying to meet the points raised by the anti-nuclear lobby, the supporters of nuclear power produce statistics of their own.

The claims of the radiation danger to the general public as a result of the minute amounts of radioactive material that are contained in the water that is discharged into the Irish Sea from Windscale are dismissed as 'grossly exaggerated'. The highest radiation dose to a member of the general public during 1979 as a result of the Windscale operations is claimed to be less than 20 per cent of the annual exposure limit recommended by the International Commission for Radiological Protection (ICRP). The small group of the general public (termed the 'critical group') who regularly consume large amounts of contaminated fish and other seafoods from the area are estimated to have received no more than 16 per cent of the ICRP dose limit in 1979. The exposure of 'typical' members of the general public who consumed fish landed in the area was very much less – no more than 1 per cent of the ICRP limit. In fact, it has been estimated that the average member of the population in the region receives, in the course of a year, no more radiation from fish caught in the Irish Sea than he would from cosmic radiation during a single jet flight across the Atlantic.

The claims regarding the build-up of plutonium on the sea bed and the possibility of it eventually forming a critical mass that could lead to a nuclear explosion are dismissed as having absolutely no basis in fact whatsoever, being typical of the ill-informed propaganda put out by the anti-nuclear lobby.

Despite anti-nuclear statements to the contrary, it is claimed that the routine emission of radioactive waste into the environment from Windscale is in fact decreasing. This is said to be partly due to the phasing of different activities, and partly due to the refurbishing of the Magnox plant. Discharges of transuranic isotopes were significantly lower in the second half of the 1970s than in the first, and discharges of caesium 137 and other fission products were significantly reduced in 1979. These reductions are expected to continue, despite the expansion of the plant to handle oxide fuel.

The hazards associated with the disposal of nuclear waste

The anti-nuclear case

In objecting to this phase of the nuclear power industry's operations, the anti-nuclear stance is very simple. They maintain that radioactive waste is dangerous, long-lived and increasing in quantity daily, and that it can *never* be safely disposed of. It is argued that the technology for total containment simply does not exist and, since some nuclear wastes remain active for many thousands of years, they will inevitably, at *some* time in the future, penetrate the containment and begin to leak into the environment, where they could contaminate the ground water, thus creating a genetic time bomb.

The prospect of being able to solidify all highly active waste before permanent burial does not seem to mollify the critics. They argue that, although scientists around the world have been working on a range of solidification processes for years, no satisfactory system has yet been found. It now seems likely that the French and British will opt for some type of glassification process, despite the fact that a number of American and Australian scientists have expressed strong reservations regarding glass as a method of storing such waste. It is claimed that the glassified waste, as an alien material, when buried in deep rock would break up easily or would leak its toxic contents when attacked by underground water.

In order to press home their objections, the various anti-nuclear groups have been very active in trying to stop test-drilling aimed at identifying suitable geological formations in which nuclear waste might be buried. In the UK, they have had considerable success, and several local authorities have flatly refused to give planning permission for these drilling operations.

Various incidents that have taken place at Windscale are often held up as examples to show that the waste management problem is an intractable one. The anti-nuclear view of the Windscale record can be presented as follows.

They maintain that the record of leaks of radioactive materials from Windscale makes alarming reading. In 1976, for example, a major leak was discovered, the radioactive contamination levels being so high that scientists have not even been able to *locate* the leak, let alone plug it. Since 1976, 100 gallons of contaminated water have been seeping into the earth from this leak *every day*.

In 1978, in one of the newer silos, hydrogen gas built up in the corroding sludge at a rate never before experienced, and reprocessing had to be halted. This gas build-up was apparently unforeseen and unexplained, and could therefore happen again at any time.

In 1979, another major leak occurred. This was by far the worst yet, resulting in no less than *30,000 curies* of radioactive material escaping into the environment.

The really worrying thing about Windscale (they maintain) is that gravel formations run under the site. Thus, the radioactive substances that are continuously leaking from the storage tanks could well get into the water that flows through these, and could quickly turn up 50 miles away, thereby polluting the drinking water of nearby towns.

A similar record of leaks from storage tanks has been observed at a number of nuclear power stations, and it is agreed that, if *temporary* storage is proving so difficult, what chance have we of succeeding in achieving *permanent* storage or disposal?

The pro-nuclear reply

Some supporters of nuclear power get very angry when such objections are voiced and say that there is probably more nonsense talked about the dangers associated with radioactive waste than about any other aspect of the nuclear power industry. A more considered response tries to put the waste disposal problem into proper perspective by exposing some of the myths supposedly created by the opposition.

Firstly, the supporters of nuclear power point out that there are two basic types of nuclear waste – short-lived, highly active waste and long-lived, low-activity waste. The former generates large amounts of heat and therefore requires cooling and shielding until the worst of its activity dies away. After ten years or so, the activity is only a small fraction of the original level, and, after a thousand years or so, is generally so low that it is *no more radioactive than the rocks from which the original uranium was mined*. There is therefore no truth in the argument that such materials will have to be guarded for 'thousands of years'.

The long-lived, low-activity waste takes longer to decay, but is normally much less radioactive, and can usually be handled safely using only plastic gloves. The main dangers from this type of waste are from inhalation or ingestion. In general, the problems associated with the disposal of such low-activity waste are very much less than those associated with the storage of highly active waste.

Secondly, they feel that it is important that people should realize the true scale of the nuclear waste problem and compare it with the waste problem associated with the burning of fossil fuels. It is not generally appreciated that the gaseous and solid waste materials discharged into the atmosphere as a result of the burning of fossil fuels such as coal and oil contain *at least* as much radioactive materials (MWh for MWh) as are released from a nuclear power station. In volume, a fossil-fuelled power station produces 3½ million times more waste than a nuclear power station of the same power, and the toxicity of the fossil fuel waste also lasts longer. If all our electricity were produced from nuclear sources, the total amount of radioactive waste would, when converted into solid form, represent a volume roughly equal to that of one pencil per person per year. On the same basis, if all the electricity consumed by the average British citizen during his/her lifetime were generated from nuclear sources, then the vitrified highly active waste would have a volume no bigger than a normal two-ounce tobacco tin (see Figure 11.6).

At present, all the highly active wastes from Britain's nuclear power programme are stored in special underground tanks at Windscale. To date, roughly 1000 cubic metres of such waste have been stored in nine of these tanks and, by the end of the century, it is expected that the total volume will be no more than 4000 cubic metres, stored in roughly 30 tanks.

Thirdly, the pro-nuclear lobby points out that there is no recorded case of anyone coming to serious harm in the UK as a result of nuclear waste. This should be compared with the large numbers of fatalities that occur every year from bronchitis and other respiratory diseases as a direct result of the atmospheric pollution produced by the burning of fossil fuels – conservatively estimated at around 20,000. The same type of comparison can be made for other countries.

The pro-nuclear lobby also has a few things to say about the much-publicized Windscale leaks. First, they point out that neither of the leaks involved the highly active storage tanks, which have *never* leaked. Leak number 1, which was first detected in 1976, involves the leakage of contaminated cooling water, *not* liquid waste, from a solid waste storage silo. This leak, it is admitted, will probably continue, albeit on a greatly reduced scale, until the transfer of the silo contents to another silo can be completed. Practically all of the activity resulting from the leak has, however, been retained in the soil close to the storage silo, well above the water table. Any radioactive isotopes not absorbed by the soil would take several years to migrate to the site boundary or into the water table, by which time their activity would have decayed to a negligible level. In their report in February 1980, the Health and Safety Executive concluded that the leak does not constitute a hazard to the safety of workers at the site, to the environment beyond the immediate vicinity of the leak, or to the general public.

Figure 11.6 *The volume of highly active waste corresponding to one person's lifetime electricity consumption (simulated)*

Leak number 2, which was discovered in 1979, involved the leakage of unconcentrated highly active waste from a building no longer in operational use. The leak was immediately stopped, and produced no significant damage to the environment according to a Health and Safety Executive report published in July 1980.

The nuclear power industry is confident that the problem of the long-term storage and disposal of highly active waste *can* be solved. France is already in a position to produce glassified waste on a commercial basis, and Britain may soon be in the same position. One long-term option would be to vitrify all highly active waste and to store it in underground repositories in stable geological formations, possibly deep formations within national boundaries.

It is not anticipated that disposal will begin before the end of this century, but *one* scheme currently under consideration would involve enclosing the glassified waste in stainless steel cans with walls 12 millimetres thick. These would be surrounded by ten centimetres of lead, and possibly by a further five millimetres of titanium, and would then be stored between 300 and 1000 metres below ground level. As stated earlier, the activity of the waste will, after 1000 years or so, have fallen to that of the original ore from which the uranium was mined, and will therefore constitute no further danger after this period has elapsed.

Claims that the radioactive contents could leak out of these storage cans or be leached out by water are said to be totally unfounded. It can be shown that the titanium outer casing would take at least 4000 years to dissolve away, even if subjected to a steady stream of salt water, while the lead inner casing should survive for even longer (the examples of Roman lead artefacts which have survived virtually unscathed at the bottom of the Mediterranean for over 2000 years can be quoted in support of this argument). Even if the radioactive materials *did* leak out of the glass, studies on the migration of the radioactive isotopes from the 'natural' nuclear reactor at Oklo in Gabon indicate that they would remain in the immediate vicinity of the store for thousands of millions of years.

Like all aspects of the nuclear power industry, the disposal of radioactive waste is controlled by law and by international conventions. The Advisory Committee on Radioactive Waste Management, which was set up in the UK following a Royal Commission recommendation, has reviewed current practices in radioactive waste management and disposal, and has reported that these are generally satisfactory and should not give rise to public concern.

The genetic dangers associated with the nuclear power industry

The anti-nuclear case

In Chapter 4, we outlined some of the possible genetic effects of radiation. The anti-nuclear lobby argues that, since the nuclear power industry is increasing the radiation dose to which individuals are subjected, it is without question adding to the potentially harmful pool of recessive genes carried by the human race, as well as giving rise to a range of immediate tragedies such as still-born births and deformed or mentally deficient children. Even very low levels of radiation are said to add to this harmful genetic load, there being no such thing as a safe dose of radiation. Also, unlike the radiation that is received naturally from cosmic rays and rocks, the radioactivity that is released into the environment by the nuclear power industry is mostly in the form of material that can be inhaled or ingested, thus doing its damage *within* the body. By allowing its gene pool to be contaminated by the radioactive materials produced by the nuclear power industry, the human race is, it is claimed, putting its long-term health and, indeed, its very existence at risk.

The pro-nuclear reply

In reply, the supporters of nuclear power readily admit that any *significant* increase in the average amount of radiation to which individuals are exposed *could* have harmful long-term genetic effects. For this reason, the nuclear power industry makes every effort to ensure that any contamination of the environment that it causes is well below the level that could produce such harmful effects. The industry strongly refutes the claim that it is putting the future of the human race at risk, and offers the following comments on genetic mutations and the extent to which their occurrence is likely to be increased as a result of its activities:

1. Mutations are not a rare occurrence, but occur all the time in all species of plants and animals. Such mutations are, in fact, widely believed to be responsible for the evolution of species, the process of natural selection causing the harmful mutations to disappear owing to the early death or infertility of the individuals concerned and enabling the rare beneficial mutations to spread throughout the species until they become universal. (Roughly 99 per cent of all mutations are thought to be harmful.)
2. Most of the mutations that occur in a species are thought to be produced by thermal excitation of the genes in the reproductive cells, there being a small but finite probability that the configuration of a gene will undergo spontaneous change as a result of thermal agitation. The remaining 10 per cent are thought to be produced by ionizing radiation, which, until comparatively recently, all originated from natural causes such as cosmic rays, radioactive rocks and radioactive body potassium and carbon.
3. From the long-term genetic point of view, the thing that matters is the average dose of radiation received by the entire population of the world over their reproductive lives; the dose received by individuals is not important. At present, the average dose is roughly three rems, and it has been estimated that this figure would have to increase roughly ten-fold in order to double the naturally occurring mutation rate. Smaller increases would increase the number of

genetic disorders, but these would have to be fairly high – of the order of hundreds or thousands of millirems – before the effect became detectable. At the moment, the average *lifetime* dose received worldwide as a result of the nuclear power industry is *less than ten millirems* – far too low to have any noticeable genetic effect.

Another way of trying to put the genetic hazard represented by the nuclear power industry into perspective is to examine the breakdown of the radiation dose that the average British citizen receives annually. This is recorded in the table in Figure 11.7, and is worthy of some comment.

		Average annual dose (millirems)	Percentage of total (rounded off so that the figures add up to 100)
Natural sources (78% of total)	Cosmic rays	31	13
	Terrestrial gamma rays	38	16
	Radioisotopes in body	37	16
	*Radon decay products	80	33
Artificial sources (22% of total)	Medical radiation	50	20.7
	Colour TV, air travel, luminous watches, etc	0.8	0.4
	Occupational exposure	0.9	0.4
	Fall-out from nuclear weapons tests	1	0.4
	Nuclear power industry	0.3	0.1
	Total	239	100

* While radon decay products in air make a large contribution to the *total* dose, they make virtually no contribution to the *genetically significant* dose, because they affect the lungs – *not* the gonads.

Source: National Radiation Protection Board (1981) *Radiological Protection Bulletin*, No 39, March

Figure 11.7 *Average annual radiation dose per person in the UK*

The first thing to observe in Figure 11.7 is that the nuclear power industry is responsible for less than a third of a millirem in an annual total of 239 millirems. That this is considered to be far too small to have any noticeable effect on the rate at which genetic disorders occur is supported by the fact that regional variations in natural background radiation dose appear to have no noticeable effect on the rate at which such disorders occur. The regional variations are, of course, *much* larger than 0.3 millirems. People living in Aberdeen, for example, receive 40 millirems a year more than those living in London. The inhabitants of other parts of the world receive even larger doses without any apparent increase in the frequency of genetic disorders (the natural background radiation dose in Kerala, India, for example, is about 400 millirems).

The table also shows that, on average, medical sources of radiation account for over 30 per cent of the total annual dose received. This artificially created increase in the average dose, which has taken place during the last 80 years or so because of the use of X-rays in medical diagnosis and treatment, does not appear to have had any effect on the rate of occurrence of genetic disorders. The amounts of radiation received during different types of X-ray examination are given in Figure 11.8, and it can be seen that they are all *very* much higher than the tiny amount received annually as a result of the activities of the nuclear power industry.

Type of examination	Dose received during examination (millirems)
Dental	20
Chest	40
Abdomen	100
Stomach	300
Lower intestinal tract	600

Figure 11.8 *Radiation doses from X-ray examinations*

Conclusions

By effectively placing the protagonists face to face, we have tried to do justice to *both* sides of the argument regarding the danger that the nuclear power industry represents to the general public. While accepting that only the staunchest of anti-nuclear groups would wish to present *all* the various arguments that we have put forward, we feel that we would be doing the anti-nuclear case less than justice if we omitted any of the more important of these, since a survey of the anti-nuclear literature ultimately reveals them all.

It is a paradoxical fact that man probably knows more about ionizing radiation than about any other environmental pollutant, but is still in considerable doubt as to the extent of the risks represented by low dose levels. At present, there is a highly sophisticated debate in progress regarding the form that the dose-response curves take at these low radiation levels. As a result, there is considerable argument as to whether or not the extra cases of cancer and genetic defects that are likely to result from low doses are really 'negligible'. This whole issue is, of course, highly germane to the social acceptability of nuclear power issue, because it is in this low-level dose range that the industry's emissions occur. If it could be proved *beyond reasonable doubt* that very low-level doses of radiation do in fact cause significant numbers of cancers and genetic disorders, then the nuclear power industry would have even greater difficulty in gaining general social acceptability.

In our opinion, there is *no evidence* to suggest that this is the case. In the low-level dose debate, the main arguments are centred on whether or not there is such a thing as a 'low-threshold dose', and, if not, whether the linear dose-response theory over- or under-estimates the radiation effects. The differences in the predictions of the various theories are so small that they make it extremely difficult to verify any one theory conclusively. The vast majority of experts in radiobiology, however, believe that the linear model (in which the number of effects produced on human health by radiation is assumed to be proportional to the dose of radiation received, for *all* levels of radiation) is nearest to what actually happens, and that the linear model is either correct or over-estimates the low-level risk. Despite this consensus, several studies carried out in recent years have implied that the linear hypothesis may be an under-estimate. The conclusions of these reports have almost invariably been found to be unconvincing, however, and, as a result of detailed analysis, one group actually subsequently withdrew their findings. (The reader is referred to the Bibliography for further details.)

Although we ourselves are not authorities on radiobiology, we are satisfied with the conclusions of the great majority of experts in the field, namely, that the tiny amount of radiation to which the general public is exposed as a result of the normal operations of the nuclear power industry constitutes no real danger either to public health or to the long-term genetic future of the human race. You, the reader, may or may not agree, but we hope that the above analysis has at least enabled you to make up your mind from a position of knowledge rather than one of ignorance or prejudice.

CHAPTER TWELVE

The Threat to Democratic Freedom and World Peace

In this final chapter, we will discuss two issues that are more closely related than many readers possibly realize, namely, the threats that nuclear power is alleged to represent to democratic freedom and to world peace.

While personal freedom can often be confused with licence, democracy and democratic freedom are more clearly understood by most people. We will therefore begin by looking at the arguments that are put for and against nuclear power in this context.

On an international scale, democracy and the proliferation of nuclear weapons become intricately linked, and there seems to be almost universal acceptance of the view that the continued expansion of national nuclear power programmes may contribute to the problem of the proliferation of nuclear weapons. What is often questioned is whether nuclear power is the prime cause of, or agent for, proliferation and whether there is anything that anyone can do to reduce the risk of such proliferation. The issues at stake are extremely complicated and, in order to try to place the subject in any sort of perspective, it will be necessary to examine both the technological and the political aspects of the problem. Clearly, it will not be possible to do the subject anything like full justice in the limited space available in a single chapter, but the interested reader can study the matter in greater depth by following up the references listed in the Bibliography.

The democratic freedom issue

The people who speak out against nuclear power argue that it not only represents a serious threat to our health, but also to the personal freedom that forms the basis of a democratic society. We are told that in Britain, for example, the nuclear power industry subjects all its employees to the totalitarian restrictions of the Official Secrets Act, forbidding them to speak about what they are doing on pain of imprisonment. In addition, the industry is accused of employing large numbers of armed police to keep the general public out of its sites. These, in effect, form a 'private army', over which, it is alleged, the public has no direct control.

If the nuclear power industry continues to expand, the anti-nuclear lobby sees these restrictions on liberty increasing – especially if we decide to go ahead with a commercial fast reactor programme. Indeed, some

American commentators claim that the widespread use of such reactors would require such stringent security precautions (because of the transportation of plutonium involved) that it would be incompatible with democracy as we know it.

It is also claimed that the nuclear power industry represents a serious internal political threat and a source of serious danger in time of war. With more and more plutonium being created and stockpiled, there is a constant risk of terrorists stealing enough to make an atom bomb, or blowing up a consignment of plutonium, with all the devastation and contamination that this would produce. Also, it is argued that nuclear power stations represent prime targets for terrorist groups or for enemies in times of war; by attacking such targets, large parts of a country could be rendered permanently uninhabitable.

The proponents of nuclear power would argue that, as in many other of the issues so far examined, the above case is riddled with misconceptions and deliberate half-truths, and that the anti-nuclear lobby is in fact guilty of malicious scare-mongering. For instance, it cannot be inferred that the freedom of British citizens is threatened in any real way just because the British nuclear power industry requires its employees to sign the Official Secrets Act and employs stringent security methods (such as its own police force) to protect its installations and property. Practically *all* industries require their employees to work under some form of secrecy, mainly to protect their commercial interests, so there is nothing unusual or unacceptable about the nuclear power industry doing the same. It is simply seen as a sensible precaution which is designed to ensure that it is the citizens of Britain who benefit from work in which they have invested large amounts of money, not some private commercial organization or foreign country. The same is true of other countries who have built up nuclear power industries and taken steps to protect these from possible industrial espionage by employing similar measures.

Similarly, it is pointed out that practically all major industries and commercial organizations now employ private security forces of one form or another in order to protect their property and interests. Anyone who has travelled by air in recent years surely appreciates that the reason for the presence of security forces at airports, for example, is to protect the freedom and rights of the travelling public. The point raised about private armies is directed specifically at the UKAEA police force, which is deployed on all UKAEA establishments. In fact, the members of this force have no more powers than any other police authorized to act in the same area, and are at all times answerable in the same way for their actions through the courts and, ultimately, through the Secretary of State for Energy and the British Parliament. As for the frequent claims that these police are 'armed' and therefore constitute a 'private army', it should be realized that firearms are only issued when fissile material such as plutonium needs to be protected at, or in transit between, establishments. This is only done to protect the general public by preventing such material from being attacked or stolen. The impression given by the anti-nuclear groups that *all* nuclear power installations are 'bristling' with armed guards is considered to be particularly mischievous. The industry quite rightly points out that the UKAEA police have no jurisdiction over the nuclear power stations that are operated by our various generating boards, and that these are not in fact protected by armed guards. Such stations have their own security forces, and the UKAEA is not involved.

There is no dispute over the point that terrorists present a serious threat to society. It is, however, pointed out that it would be very difficult indeed for terrorists to acquire the materials needed to make an atom bomb, and practically impossible for them to manufacture one. All fissile material, in whatever form, is carefully guarded at all times, both in store and in transit and, in any case, is not normally in a form suitable for the manufacture of bombs. Plutonium, for example, would first have to be purified, something that would be virtually impossible to do, since the process would have to be carried out in a specially designed reprocessing plant. Also, during most stages in the fuel cycle, the fissile material is in a highly radioactive form, and would therefore be very dangerous for terrorists to handle. The loss of fissile materials from nuclear plants is safeguarded against by the most stringent security and accounting systems. These are designed to ensure that no fissile material can ever be stolen, and that no significant accidental losses can occur.

As to possible terrorist attacks on nuclear installations, it is argued that these would be extremely difficult to mount because of the security measures that are taken to protect such plants. Also, such attacks would be unlikely to cause any major hazard to the general public *even if successful* unless they took place on a truly massive scale. Terrorists could, it is said, do much more damage by attacking such things as chemical plants or oil stores. Similarly, any enemy country wishing to attack a country like Britain could do far more damage by attacking her offshore oil installations, oil refineries and chemical plants than it could ever do by attacking her nuclear power stations. There are, of course, contingency plans for dealing with all these situations, but it is obviously not in the public interest to make these generally known.

In today's highly complex society, it seems inevitable that we must be prepared to accept *some* measure of restriction on personal freedom and access to information in order to protect our collective interests. This is, of course, not a new situation, and seems likely to remain with us for the foreseeable future, but we see no reason why it should be regarded as a serious threat to democratic freedom provided that it is not taken to extremes.

The nuclear proliferation issue

The danger of international proliferation of nuclear weapons has possibly now become *the* main issue influencing the future growth of nuclear power throughout the world. Although the risks of nuclear proliferation have been recognized since the earliest days of nuclear power, President

Carter's 1977 initiative in imposing a moratorium on the reprocessing of nuclear fuels and, in particular, on the recycling of plutonium, has stimulated world interest in the proliferation issue.

The principal fear, which is shared by both the anti-nuclear and pro-nuclear factions, is that the continued expansion of the civil nuclear power programme into more and more parts of the world will inevitably increase the risk of 'politically unstable' regimes being able to acquire a nuclear weapons capability, something which could result in a serious threat to world peace. It is feared that such regimes could divert fissile materials from their civil nuclear power programmes and use them to build nuclear weapons.

Where the differences of opinion occur is in the potential role identified for civil nuclear power in such proliferation. On the one hand, it is argued that there are other, easier ways for a country to acquire the fissile materials it needs other than through building nuclear power stations, and that this route is not one of the main proliferation dangers. The opposite view stresses that, since the civil nuclear power programme is the principal source of plutonium, and since plutonium can be used in the manufacture of nuclear weapons, it is essential that the strictest controls are kept on the 'availability' of plutonium; thus, any further spread of nuclear power technology can only increase the risk of nuclear proliferation and should therefore be avoided.

INFCE

As a result of the renewed concern about nuclear proliferation that stemmed from President Carter's measures, an international forum was set up in October 1977 – the *International Fuel Cycle Evaluation* programme (*INFCE*). Participation in INFCE was open to all interested states and relevant international bodies. In the event, 66 countries and five international organizations took part, representing industrialized and developing countries with nuclear power and nuclear weapons programmes at various stages of advancement and also consumers and suppliers of materials, technology and equipment. INFCE's function was to undertake a technical evaluation of different fuel cycles in the context of possible future world nuclear power developments. Its main aim was to determine how the proliferation risks of the wider use of nuclear power might be reduced, *not* to assess the desirability or otherwise of nuclear power as compared with other sources of energy. (It had already been agreed by the participating nations and bodies, at the Organizing Conference, that nuclear power was essential.)

To practically no one's surprise, INFCE did not come up with a technological solution to the proliferation problem (There simply isn't one!) Its conclusions tended to be general rather than specific, although it was agreed that the problem was basically a political one. It was felt that the problem of proliferation of nuclear weapons would not simply disappear by banning the construction of nuclear power stations. Nuclear weapons had, after all, been in existence long before any commercial nuclear power stations were built; and, just as the wheel cannot be uninvented, neither can nuclear power.

The principal conclusions from INFCE can be summarized as follows:

1. The strength of feeling among those who took part on the continued need for nuclear power should not be underestimated. The underdeveloped countries, in particular, insisted that their future energy needs should be taken into account.
2. Most countries wished to keep their options open regarding reprocessing and the use of fast reactors because:
   (a) there is likely to be a possible squeeze on uranium supplies at the beginning of the next century,
   (b) there is a severe regional disparity in uranium reserves,
   (c) the potential role of the fast reactor and the timing of its introduction could alleviate future supply problems.
3. Proliferation was seen as a *political* problem, and the solution did not appear to be in postponing decisions about the introduction of particular fuel cycles such as those that use plutonium, since:
   (a) there is no 'technifix' solution or fuel cycle that completely avoids the use of potential weapons material,
   (b) construction and subsequent misuse of fuel cycle facilities is not the easiest or most efficient route to the production of weapon-grade fissile material,
   (c) it is extremely difficult to distinguish between different fuel cycles in terms of their respective proliferation risks.
4. It was agreed that there was a need to obtain consensus on the non-proliferation conditions that should be applied to nuclear trade, since any uncertainty over the reliability or availability of nuclear fuel supplies could lead countries to seek nuclear independence earlier than would otherwise be necessary.

INFCE did not (as is sometimes claimed) represent a defeat for US policy, although it is fair to say that its conclusions generally support the British and French view on nuclear power, which differs significantly from that of the USA. The form of the recommendations has, however, given the critics of US nuclear policy further ammunition. Indeed, some cynics argue that there was an ulterior motive behind the Carter initiative. They allege that the Americans, realizing that they had fallen behind countries such as France, Russia and Britain in areas such as fast reactor technology, tried to adopt a 'dog in the manger' attitude rather than giving a lead to the rest of the world. Some critics even accuse the Americans of attempting to resort to political and economic blackmail. (The Americans have, of course, been accused of this before – over the issue of supersonic transport, for example. Here, America abandoned her own highly ambitious SST project on the grounds of technical difficulties and escalating costs, leaving the field open to the British/French Concorde. The latter has, however, never achieved commercial success – partly, it is argued, because of American obstruction.)

We feel that this analysis of the American position is far too harsh, and does not take account of several important factors. Firstly, as we have shown in Chapter 7, the future energy supply position seems much brighter for the USA than for other nuclear countries such as Britain and France. The USA therefore does not need to expand its nuclear power programme so rapidly. Secondly, the anti-nuclear lobby in America is very powerful and well organized, and has influence at the highest levels in government. Thirdly, America has indigenous supplies of uranium that are currently under-exploited, and is therefore not under the same pressure as Britain and the rest of Western Europe to develop fast reactor technology.

The Carter policy does, however, have a number of unfortunate side effects, some of which could, paradoxically, actually *increase* the risks of nuclear proliferation. It results, for example, in an increase in total demand for uranium; it also increases world reliance on the USA for enriched fuel, and helps to widen the gap between those who 'have' and those who 'have not'. What is good for one nation is not necessarily good for another, and herein lies the root of the problem.

Technical aspects of proliferation

Although it is clear that there can be no purely technological solution to the proliferation problem, this does not mean that technology can be of no use in reducing the risks of proliferation. Most of the concern is centred around the so-called 'plutonium economy', and is therefore intimately connected with the future role of the fast reactor. We have already stated that the current emphasis on plutonium in the proliferation debate tends to distract people from what are perhaps the more serious risks associated with other fissile materials such as uranium 235. Supplies of weapon-grade U235 can be produced *independently of any reactor programme* provided that a suitable uranium enrichment plant is available. Thus, it could be argued that there is a far greater risk of nuclear proliferation taking place as a result of the spread of uranium enrichment plants – which are now much easier and cheaper to build than was previously the case – than there is as a result of the spread in the use of plutonium. As we have seen, the type of plutonium that is produced as a by-product of the operation of thermal reactors, or is bred in fast reactors, is totally unsuitable for use in the manufacture of nuclear weapons unless it is first purified – something that can only be done in a highly sophisticated, highly expensive plant.

Nevertheless, the possibility of plutonium being obtained from some sort of reactor programme, purified and used in the manufacture of nuclear weapons does undoubtedly exist – and does constitute a potential threat to world peace. In order to reduce the proliferation risks from plutonium, it is therefore necessary to reduce the amount of plutonium that is 'available'. It has been estimated that there are currently over 75 tonnes of plutonium in non-nuclear-weapon states, all subject to strict safeguards, plus an unknown amount in the various nuclear-weapon states. Of this, roughly 11 tonnes is in separated form, the remainder being contained in spent fuel elements and in reactors. This total figure is increasing at about ten tonnes per year. *What often seems to be forgotten is that, if our future nuclear power programme is based solely on thermal reactors, the amount of plutonium created will continue to increase whether or not the expansion of the programme is halted and whether or not the spent fuel is reprocessed.*

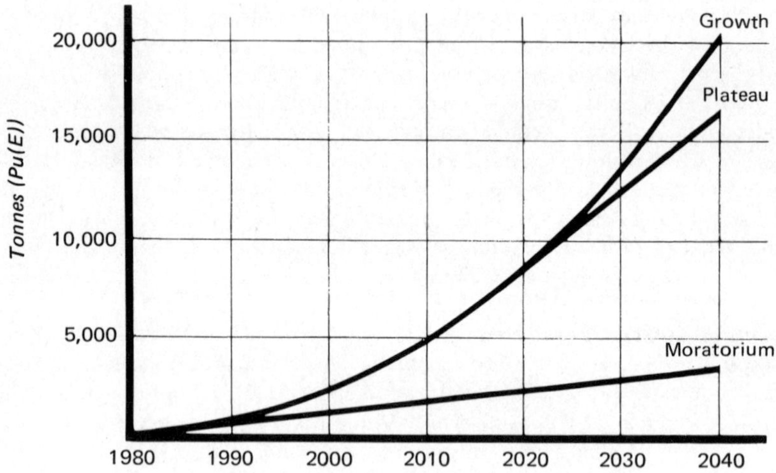

Figure 12.1 *The total future plutonium inventory from different scenarios based on the use of thermal reactors*

If we consider the American suggestion that all nations should adopt the once-through fuel cycle, the future implications for the total inventory of plutonium can be seen in Figure 12.1. Here, three different scenarios for the future of nuclear power have been chosen: (i) A *moratorium*, in which world nuclear capacity remains constant after about 1985, (ii) a *growth* in world nuclear capacity which is equivalent to the historic growth rate to date, and (iii) a *plateau* scenario, in which world nuclear capacity levels off after the turn of the century. In each case, the amount of plutonium is seen to be increasing year by year *into the middle of the next century*. Thus, not only would the once-through fuel cycle increase the risks of proliferation by increasing the demand for uranium and therefore encouraging the 'have-not' nations to go it alone with independent reprocessing and enrichment facilities, but the total amount of plutonium in the world would actually continue to increase as a result. We can therefore see that widespread adoption of the American policy would probably lead to a significant *increase* in the risk of proliferation – not a *reduction*, as is claimed by its protagonists.

This paradox can only be resolved if world plutonium stocks can be reduced in size. *At present, the only way in which plutonium can be destroyed is to use it as the fuel for fast reactors.* In this way, not only could the amount of plutonium in the world be reduced, but it could also be made very much less 'accessible' for diversion into nuclear weapons programmes. Let us now consider how this might be achieved.

First of all, we should remember that the word 'fast' refers to the speed of the neutrons that are used to produce fission in a fast reactor, *not* to the speed at which such a reactor breeds plutonium. Fast reactors are in fact rather *slow* breeders of plutonium – slower than many thermal reactors. A typical thermal reactor (a 1000-MW PWR) produces roughly 300 kilogrammes of plutonium each full operating year. A typical fast reactor of the same generating capacity which has a breeder blanket around the core will incinerate plutonium in the core and produce plutonium in the blanket. A typical rate of incineration is about 200 kilogrammes per full operating year and, depending on the particular blanket design, the rate of production of plutonium can be as high as about 450 kilogrammes. Thus, in this case, we would have a net rate of production of plutonium of about 250 kilogrammes a year. By altering the design of the blanket region, the reactor as a whole can be operated either as a net *producer* or as a net *incinerator* of plutonium. This ability to be a net incinerator of plutonium is crucial to the proliferation issue, and is one of the outstanding features of the fast reactor.

While a fast reactor is in operation, the plutonium contained within it is clearly 'inaccessible'. Thus, if all the plutonium outside thermal reactors were inside fast reactors, then the risks of proliferation from the 'misuse' of plutonium would virtually disappear. This cannot ever be the case, of course, since it is necessary to reprocess the spent fuel elements of thermal reactors (and fast reactors) in order to extract the plutonium, and then to fabricate that plutonium into new fast reactor fuel elements. It is this intermediate stage of recovering plutonium from spent fuel elements and then using it to make fresh fuel that has given rise to the concern about the so-called 'plutonium economy' that was mentioned earlier.

It should be remembered that it is not possible to establish a large fast reactor programme overnight. In the early stages of a fast reactor programme, a viable thermal reactor programme would still be necessary in order to sustain an increasing number of fast reactors until the required number of the latter had been established (see Chapter 9). The proliferation risks would therefore not disappear for some considerable time.

The launching of a fast reactor programme thus appears to offer no immediate reduction in the proliferation risk. Once plutonium has been isolated from a spent fuel element, it is no longer 'inaccessible' and, until it is fed back into a fast reactor, it is available for misuse. Use of fast reactors is therefore unlikely to reduce the proliferation risks during this 'launch' phase of their operation, although it must also be said that it is unlikely to increase them either; any plutonium that is fed into a fast reactor immediately becomes 'inaccessible', and therefore effectively reduces the world's stock of potential weapon-grade materials.

The fuel cycle that will be needed to *maintain* fast reactors once a programme of the required size has been established, with the aid of thermal reactors, will be very different from the launch cycle discussed above. There are, for example, good reasons to suppose that the 'turn-round time' for plutonium will be very much shorter than for the launch cycle, and so it will be possible to retain the plutonium in a far less 'accessible' form for a much longer proportion of the time for which it exists. It is anticipated, for instance, that it will be possible to reprocess the fuel in something like a year after it has been removed from a reactor, rather than the ten to 20 years that is the present practice for some thermal reactor fuels. This means that the fuel elements will still be highly radioactive, and that the plutonium in the reprocessing plant will remain virtually 'inaccessible' throughout the entire process. This 'inaccessibility' could be greatly increased by deliberately leaving some of the highly active fission products in the plutonium after reprocessing. By adopting such measures, it should be possible to have a total turn-around time of under two years for the plutonium used as fast reactor fuel, and to reduce to a very low level the amount that is available or 'extractable' for misuse at any given time.

The amounts involved would, of course, depend on the future energy policy that was adopted, and a range of possible scenarios is shown in Figure 12.2. Here, it is assumed that the phasing in of fast reactors will begin in the 1990s, and that they will gradually replace thermal reactors during the period to the middle of the next century. Since it is the amount of *'extractable'* plutonium that causes concern, it can be seen that the fast reactor and its associated technology could play a major role in *reducing* future proliferation risks.

Political aspects of proliferation

No amount of technological innovation can safeguard the world from the dangers of nuclear proliferation. As the INFCE concluded, it is essential that all countries should work towards international understanding and cooperation if the dangers of proliferation are to be minimized. In a sense, we feel that this pronouncement was designed to remind the participating nations, and the rest of the world, that channels *already* exist through which it should be possible to build towards international understanding and cooperation.

For instance, the International Atomic Energy Agency (IAEA), which was established in 1957, is a United Nations Agency that has 109 member states. Its basic objectives are to enable nuclear energy to make the maximum contribution to world energy requirements and, at the same time, to prevent the spread of nuclear weapons. In pursuit of this second objective, the IAEA has constantly stressed that countries which are prepared to sign and ratify the Non-Proliferation Treaty (NPT) and place their nuclear facilities under the Agency's safeguards and agreements should be given full support in their nuclear power programmes through

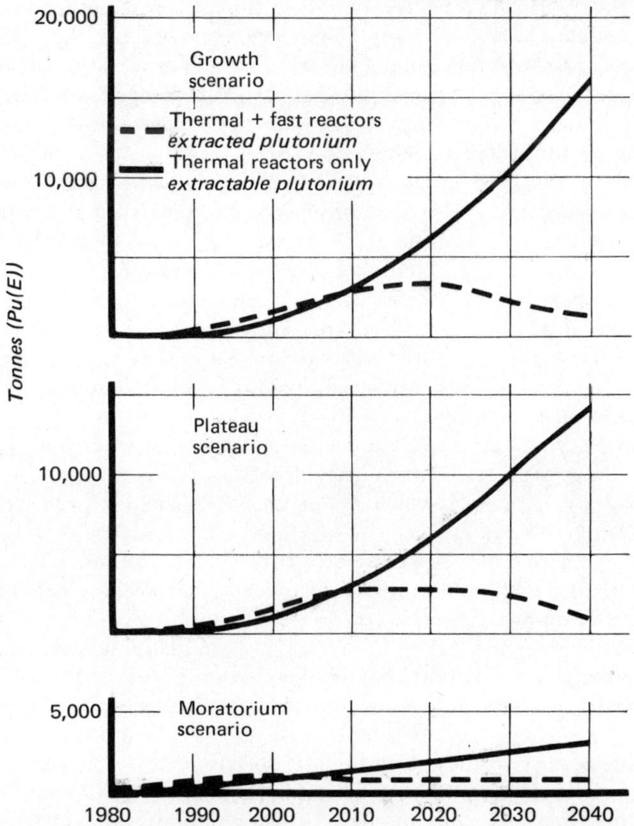

Figure 12.2 *The likely effect of fast reactors on the amount of 'available' or 'extractable' plutonium available at any given time*

the supply of technology and plant from other member countries. Although the USA and Britain, as established nuclear-weapon states, are not required, under the terms of the NPT, to place their non-military nuclear programmes under IAEA safeguards, they have in fact chosen to do so voluntarily. The basic steps in applying safeguard procedures involve periodic design reviews, the maintenance of nuclear plant records and reports, and regular inspections, all of which have implications for national security and sovereignty.

The Non-Proliferation Treaty was also sponsored by the United Nations when it was endorsed by the UN General Assembly in 1968. To date, 103 countries have signed and ratified the Treaty (of which 100 are non-nuclear-weapon states) and a further 11 countries have signed but not yet ratified it. Basically, the Treaty obliges the nuclear-weapon powers *not* to supply nuclear weapons to any non-nuclear-weapon state, or in any way to assist such a state to develop a nuclear weapons capability. At the same time, the NPT does not restrict any party to the Treaty from developing and using nuclear power for *peaceful* purposes. It does, however, require that all parties to the Treaty should make strenuous efforts, by negotiation, to bring about measures which will eventually result in the cessation of the nuclear arms race and lead to general nuclear disarmament under strict and effective international control.

While no one would claim that the IAEA and the NPT provide a guarantee against nuclear proliferation, it can be said that they represent a major step along the road. Ideally, of course, every country in the world should be members of both, but, as we have seen, membership has implications for national sovereignty, and this has undoubtedly acted as a disincentive in some cases.

In Europe, the Euratom Treaty imposes additional constraints on the signatory European states through its own system of safeguards and inspections. Even France, which is not an NPT signatory, has agreed to the Euratom access rules.

These international co-operative efforts to control the spread of nuclear weapons surely *must* form the basis of our efforts to reduce proliferation risks to a minimum. They have, of course, been criticized for not working properly, and for being full of loopholes. No one really disputes this, but most commentators feel that the only way to move forward is to develop these systems of international control through political and technological initiatives.

While in no way wishing to cast doubt on President Carter's sincerity or integrity, we ourselves feel that his attempt to stop the development of fast reactors was ill-advised, both from a technological and from a political point of view. Instead of stopping the use of fast reactors, we would like to see those countries who already have fast reactor programes at a prototype stage proceed with their development and exploitation as quickly as possible, *under careful international supervision*, in order to prevent the world's stocks of 'accessible' or 'extractable' plutonium from increasing in the way shown in Figure 12.1. In order to do this, and, at the same time, to allow emerging nuclear nations to continue with the programmes that are so vital for their development, it will be necessary for the major nuclear countries to take on most of the responsibility for the reprocessing of spent thermal reactor fuel and for the supply of new fuel. This would greatly restrict the spread of the 'plutonium economy', so that international supervision of *all* plutonium operations would have a good chance of being achieved.

Irrespective of the introduction of the fast reactor, the spread of conventional reprocessing and enrichment technology should, we believe, be discouraged, or should, at the very least, be controlled by international agreement. Such measures would place a heavy responsibility on the shoulders of the reprocessing nations and the international agencies in order to make sure that their role as 'keepers' of the world's nuclear fuel was not abused. Placing all reprocessing and enrichment facilities under international control would, we believe, greatly reduce the risks of proliferation – not just from plutonium, but also from other fissile materials such as U235. Certainly, we are convinced that the answer to the proliferation problem does not lie in placing artificial restrictions on the spread of nuclear power, which the developing world rightly regards as one of the keys to its future prosperity. Denying the benefits of nuclear power to any nation that can be seen to have a need for it would, in any case, probably *increase* the danger of proliferation, by forcing that nation to 'go it alone'. Whether we like it or not, nuclear power is almost certainly here to stay. Let us therefore try to ensure that we use it for the benefit of mankind, and not as a means of bringing about his eventual extinction.

Conclusion

These, then, are the various arguments that form the basis of what has become known as 'the nuclear debate'. Let us now see what conclusions we can draw from them, dealing in turn with the three basic issues examined in Parts II, III and IV above.

1. Do we really need nuclear power?

In Part II, we tried to answer this question by taking a detailed look at the recent energy consumption pattern of the Western world, examining possible future trends, and drawing up an inventory of the various non-nuclear energy sources that are likely to be available by the year 2000 – including various 'alternative' sources of energy currently under development. This analysis made it clear that the USA, with its vast untapped reserves of coal, seems likely to be able to support a *modest* increase in energy consumption by the end of the century without making significant use of nuclear power, but nothing like enough to allow the American economy to expand at the sort of rate her citizens have come to expect. Other Western countries such as Britain, on the other hand, appear to be in a much weaker position and, unless they make much greater use of nuclear power, will not be able to sustain *any* growth in energy consumption. Indeed, those countries which lack indigenous supplies of fossil fuels could well have to *reduce* their total energy consumption.

Since the situation seems almost certain to get progressively worse after the turn of the century, as the world's supplies of easily accessible oil start to run out, the only realistic conclusion we can come to is that the Western world will have to make much more extensive use of nuclear power if it is to sustain the systematic increase in total energy consumption that has enabled the standard of living of its citizens to rise steadily since the end of the Second World War. The only alternative would be to adopt a radically different economic and industrial policy, together with a drastic change in lifestyle, that would, we suggest, be totally unacceptable to the great majority of people.

2. What sort of nuclear power policy should we adopt?

This question was discussed in Part III, where we first looked at the development of nuclear power to date and then examined the possible contribution that it could make to meeting our future energy requirements. Three important points emerged from this analysis.

Firstly, we saw that nuclear power is now extremely attractive from an economic point of view, and seems almost certain to become even more attractive if (as seems inevitable) the price of fossil fuels continues to rise at greater than the general rate of inflation. Thus, it could be argued on *purely economic grounds* that we should make much more extensive use of nuclear power than we do at present.

Secondly, we saw that it will not, unfortunately, be possible for the world to adopt an indefinite policy of continued nuclear expansion based solely on the use of thermal reactors. The world's reserves of economically recoverable uranium are comparatively small and, if the present expansion of nuclear power continues, seem likely to run out sometime during the third or fourth decade of the 21st century.

Thirdly, we saw that it should be possible to avert such a uranium shortage by developing commercial fast reactors, which would make much more efficient use of uranium than thermal reactors. By embarking on a properly balanced 'mixed' programme of building both thermal and fast reactors, and gradually replacing the former with the latter, we should be able to make extensive use of nuclear power well into the next millenium – or until some alternative source of energy such as nuclear fusion or solar power can solve our long-term energy problems.

3. Is nuclear power socially and environmentally acceptable?

This final question, which is to most people *the* key issue regarding nuclear power, was discussed in Part IV, where it was tackled in three stages in Chapters 10, 11 and 12. In each, the procedure adopted was to examine the arguments that are commonly put forward by the opponents of nuclear power on the one hand, and by its supporters on the other, so that a balanced view could be obtained.

In Chapter 10, we tried to determine whether workers in the nuclear power industry are subjected to unacceptable health and/or accident risks. On the basis of the available evidence, we ourselves are fairly satisfied that, *provided the agreed regulations and procedures are rigidly adhered to*, nuclear workers are not put at special risk. From personal experience, we know that the British nuclear power industry maintains the highest possible standards regarding such matters and, as a result, has an excellent health and safety record. Unfortunately, not every country appears to have maintained the same high standards, reports of 'incidents' which have taken place in certain privately operated plants in the USA giving us particular cause for concern. However, a combination of adverse publicity and pressure from anti-nuclear groups and workers' organizations has led to the situation being improved.

In Chapter 11, we looked at the hazards that the nuclear power industry represents to the general public. Here, we are satisfied that there is no real cause for public concern. The risk of a major accident that causes injury or death to members of the general public is extremely small and, *on the basis of any objective comparative risk assessment*, is (we believe) socially acceptable. The same is true of the risks to the overall health of the general public (through increased incidence of cancer), and to the genetic future of mankind, that are likely to result from the normal activities of the nuclear power industry. In neither case is the tiny contribution that the industry makes to the normal background radiation level likely to have any noticeable effect.

Finally, in Chapter 12, we looked at the dangers that the nuclear power industry represents to democratic freedom and (by increasing the risk of nuclear proliferation) to world peace. On the first of these issues, we believe that there is simply no case for the industry to answer, since its activities do not reduce the liberties of the average citizen in any meaningful way. On the second, we believe that there is rather more cause for concern, since the present *laissez-faire* policy of unrestricted nuclear expansion does, we believe, increase the risk of the further spread of nuclear weapons. However, we are firmly convinced that the solution to the proliferation problem should be sought through international agreement and *not* by placing artificial restrictions on the growth of nuclear power in the developing world. *Provided that the spread of nuclear power is subject to responsible international control*, we believe that the risks of proliferation could be kept to a minimum.

In conclusion, we would like to make a few observations on the role of pressure groups (such as the various anti-nuclear pressure groups) in an 'open' democratic society. In general, we believe that such groups fulfil an extremely valuable social function – by subjecting the activities of central and local government, industry and commerce to continuous critical scrutiny and, in many cases, preventing or minimizing the effects of socially undesirable developments. However, we believe that such groups can, in some cases, exert such a strong influence on public opinion that they actually *prevent* socially desirable developments from taking place. This has certainly been the case over the issue of the fluoridation of public water supplies, where, in Britain and many other countries, the activities of a few well-organized anti-fluoride pressure groups have succeeded in holding back what virtually the entire medical profession regards as a wholly desirable preventative measure that could bring about a dramatic improvement in dental health.

We believe that there is a very real danger of the same thing happening over nuclear power, and that the activities of anti-nuclear pressure groups could well prevent the Western world from reaping the benefits of the one source of energy that could guarantee its continuing prosperity. This, in our opinion, would be nothing less than a catastrophe. We hope that this book will help to ensure that common sense prevails, and that nuclear power will be allowed to fulfil the role of which it is capable.

Glossary of Terms relating to Nuclear Power

A

active: a term often used as a synonym for *radioactive*

active area: see *radiation area*

activity: the number of nuclear disintegrations occurring per unit of time in a quantity of radioactive material; measured in *becquerels* (Bq) or *curies* (Ci) (see p 59)

AEA: see *UKAEA*

AEC: see *USAEC*

AECL: *A*tomic *E*nergy of *C*anada *L*td

AGR: *A*dvanced *G*as-cooled *R*eactor: the successor to the Magnox system in the UK nuclear power programme (see pp 29 and 123)

alpha decay, emission: a type of radioactive decay in which an unstable heavy nucleus emits an *alpha particle* (see p 60)

alpha particle: the particle emitted during *alpha emission*, consisting of two protons plus two neutrons, ie a helium nucleus (see p 60)

artificial element: an element that does not occur in nature, but can be manufactured from another element, eg plutonium

artificial radioactivity: see *induced radioactivity*

atom: the smallest particle of an element which has the chemical properties of that element (see p 15)

atom bomb: a popular (although incorrect) term for a *fission bomb* (see p 20)

atomic nucleus: the small positively charged central body in an atom that carries virtually all of its mass (see p 16)

atomic number: the number of protons in the nuclei of the atoms of a given element

atomic pile, reactor: outdated names for a *nuclear reactor*

B

background radiation: the radiation that is produced by natural sources such as cosmic rays and radioactive materials in rocks (see p 173)

base load: in electricity generation, the minimum steady power demand on the system over a period; normally met by stations of the highest *merit order*

base load plant: an electricity generating station that is run continuously in order to supply part of the *base load*

becquerel (Bq): the new unit of activity in the SI System; it is equivalent to 1 disintegration per second, or roughly 2.7×10^{-11} Ci

BEIR: Committee of the US National Academy of Sciences on the *B*iological *E*ffects of *I*onizing *R*adiation

beta decay, emission: a type of radioactive decay in which an unstable nucleus emits a *beta particle* (see p 61)

beta particle: the particle emitted in beta decay – either an *electron* (β^-) or a *positron* (β^+)

binding energy: the energy needed to separate a nucleus into its constituent protons and neutrons, ie the energy tied up in holding the nucleus together (see p 18)
biological shield: the layer of material (usually concrete) that surrounds a nuclear reactor or other radiation source in order to protect the operating personnel from ionizing radiation (see p 28)
blanket: a layer of fertile material (usually depleted uranium) placed round the power-producing core of a fast reactor in order to produce fissile material such as plutonium (see p 35). In fusion research, the neutron-absorbing lining of a torus (see p 43)
BNFL: *B*ritish *N*uclear *F*uels *L*td (the State-owned company responsible for the manufacture and reprocessing of nuclear fuel in the UK)
body burden: the amount of radioactive or toxic material in a human or animal body at any given time
boron: a chemical element that strongly absorbs neutrons; used in the manufacture of *control rods* and *shut-down rods* for nuclear reactors
breeder blanket: see *blanket*
breeder reactor: a nuclear reactor which produces more fissile material than it consumes (see p 37)
burn-up, burn-up factor: a measure of the percentage of fissile material in a fuel element that is burned in a given irradiation cycle (see p 44); the term *burn-up* is also applied to the heat output (in MW days per tonne) of nuclear fuel over a given period
BWR: *B*oiling *W*ater *R*eactor – a type of light water reactor in which steam is generated directly in the reactor core by letting the coolant boil (see p 33)

C

cadmium: a metal that strongly absorbs neutrons; often incorporated into steel to produce cadmium steel – a material used in the manufacture of control rods and shut-down rods for nuclear reactors
calandria: a closed tank penetrated by pipes so that liquids in each do not mix; used in the *CANDU* and *SGHWR* reactor systems (see p 33)
calorific value: a measure of the amount of heat energy that is released when a given quantity of fuel is completely burned
can: the container (usually of metal) in which nuclear fuel is sealed to prevent contact with the coolant or escape of fission products, etc (see p 27)
CANDU reactor: a type of thermal reactor developed in Canada (see p 33)
capture: the process in which a particle (eg a neutron) collides with a nucleus and is absorbed by it
carbon dioxide: a gas used as a coolant in graphite-moderated thermal reactors such as the *Magnox* and *AGR* systems
cascade process: a progressive sequence of operations in which the process material flows from one stage of the plant to the next, eg in *solvent extraction* or *enrichment* plants (see p 47)
cave: a heavily shielded compartment in which highly radioactive materials can be safely kept, handled or examined by remote manipulation (see p 53)
CDFR: *C*ommercial *D*emonstration *F*ast *R*eactor (see p 138)
CEA: *C*ommissariat à l'*E*nergie *A*tomique; the French atomic energy research organization
centrifuge: see *gas centrifuge*
ceramic fuel: nuclear fuel in the form of a hard, heat-resistant ceramic material such as uranium oxide or carbide (see p 29)
CERN: European Organization for Nuclear Research (based in Geneva)
chain reaction: a process which, once started, provides the conditions for its own maintenance, eg the fission chain reaction that takes place in a nuclear explosion or nuclear reactor (see p 25)
charge: the total amount of fuel placed in a nuclear reactor; to place fuel in a reactor

charge/discharge machine: a mechanical device for putting new fuel into or removing spent fuel from a nuclear reactor (see p 49)
charging machine: see *charge/discharge machine*
CHP: combined heat and power
cladding: a protective layer (usually of metal) separating chemically active components, eg a fuel can separating reactor fuel and coolant (see p 27)
combined heat and power: the combination of heat and power generation within the same plant (see p 94)
commercial reactor: a power-producing reactor built by a generating board or power company in order to produce electricity for commercial sale
commissioning: the process of bringing a plant (such as a nuclear reactor) into service by running it up to power, checking that it complies with its design specification, and so on
compound: a substance produced by the chemical combination of two or more *elements* (see p 16)
compound nucleus: a highly excited nucleus of short lifetime formed as an intermediate stage in a nuclear reaction (see pp 21 and 23)
containment: the prevention of release of radioactive materials from a nuclear reactor, reprocessing plant, etc (see p 156); in nuclear fusion research, the system used to contain the plasma (see p 39)
containment time: in nuclear fusion research, the average time for which a particle remains trapped in a plasma containment system (see p 39)
contaminated zone: an area that has become contaminated with radioactivity
contamination (radioactive): the spread of radioactivity to places where it should not be (eg to clothing, work surfaces, etc)
controlled fission: steady, stable fission of the type that takes place in the core of a nuclear reactor (see p 25)
control rods: rods of neutron-absorbing material such as boron or cadmium-steel that are moved into and out of a reactor core in order to control the power level (see p 27)
conversion: nuclear transformation of *fertile material* into *fissile material* by neutron absorption (eg in the core of a nuclear reactor – see p 35)
coolant: the gas, water or liquid metal that is circulated through the core of a nuclear reactor in order to remove the heat generated by the fission of its fuel (see p 28)
coolant circuit: the circuit through which the coolant flows between the core of a nuclear reactor and the heat exchangers or turbines where it gives up its heat (see p 28)
cooling off: the process whereby spent fuel elements are stored under water after being removed from a nuclear reactor in order to allow the worst of their radioactivity to die away and the associated heat to be dissipated safely (see p 50)
cooling pond: a pond in which spent fuel elements are stored for the purpose of *cooling off*
core: the central region of a nuclear reactor containing the fuel elements where the fission chain reaction takes place
cosmic rays: radiation incident on the earth from space comprising X-rays, gamma rays, neutrons, protons and other charged particles plus the secondary radiation produced when the original particles interact with the earth's atmosphere
critical: a nuclear reactor or other assembly of fissile material is said to have 'gone critical' when its chain reaction has just become self-sustaining
criticality incident, excursion: inadvertent accumulation of fissile material into a critical assembly leading to criticality and the sudden and dangerous emission of neutrons, gamma rays and heat
critical mass: the minimum mass of a given type of fissile material which can become critical in a particular configuration
CTR: *c*ontrolled *t*hermonuclear *r*eaction (see p 38)
curie (Ci): the most commonly used unit of activity, corresponding to 3.7×10^{10} disintegrations per second (see p 59); currently in the process of being replaced by the *becquerel*

D

de-canning: removing the canning material from a fuel element prior to reprocessing (see p 53)
decay: see *radioactive decay*
decay product: the nucleus that is produced as a result of radioactive decay
decommissioning: the final closing down and putting into a state of safety of a nuclear reactor or other plant when it has come to the end of its useful life
decontamination: removal of radioactive contamination
depleted uranium: uranium with less than the natural proportion (0.7 per cent) of U235, eg the residue from an enrichment plant or reprocessing plant (see p 47)
design life: the time for which a nuclear reactor or other plant is expected to remain operational
deuterium: the hydrogen isotope of mass number 2; often described as *heavy hydrogen* (see p 18)
DFR: *D*ounreay *F*ast *R*eactor (see p 124)
diffusion plant: a plant in which enriched uranium is produced by the gaseous diffusion process (see p 47)
disintegration: any transformation of a nucleus, either spontaneous or induced, in which particles or gamma rays are emitted
disposal: the removal from man's environment of unwanted or dangerous material such as nuclear waste (see p 54)
dose: a quantity of radiation; often used as an abbreviation for absorbed dose or dose equivalent
dose limit: the maximum dose to which a person or organ is allowed to be exposed over a given period of time (see p 142)
dose rate: the dose absorbed in a given unit of time, eg rems per year
doubling time: in a breeder reactor, the time taken to produce enough surplus fissile fuel to provide the initial charge for a new reactor of the same size (see p 57)
dumping: disposing of nuclear waste at sea under carefully controlled conditions (see p 55)

E

economic reserves: the reserves of a resource such as uranium which it is economic to exploit at a given time (see p 96); also known as *recoverable reserves*
EDF: *E*lectricité *d*e *F*rance; the French electricity-generating organization
effluent: solid, liquid or gaseous waste discharged from a plant or process into the environment
electrical output: the electric power produced by a generating plant such as a nuclear power station – usually measured in *megawatts*; also, the integrated output over a period of time – usually measured in megawatt hours or megawatt days
electromagnetic radiation: radiation consisting of electromagnetic waves; in the nuclear context, it includes gamma rays and X-rays
electron: one of the stable sub-atomic particles of which all matter consists, carrying a single unit of negative electric charge (1.6×10^{-19} coulombs) and having a mass of 9.1×10^{-31} kilogrammes (see p 16)
electron volt (eV): a unit in which energy is measured in the study of sub-atomic particles and their interactions, being equivalent to the work done when one unit of electronic charge moves through a potential difference of one volt (see p 19)
element: a substance which cannot be broken down into simpler components by chemical means (see p 15)
emergency shut-down system: a system for shutting down a reactor quickly in case of emergency, eg by dropping neutron-absorbing shut-down rods into the core (see p 27)
enriched uranium: uranium in which the proportion of U235 has been artificially increased above its natural proportion of 0.7 per cent (see p 46)
enrichment plant: a plant in which enriched uranium is produced – generally by *gaseous diffusion* or using *gas centrifuges* (see p 47)

ERDA: *E*nergy *R*esearch and *D*evelopment *A*dministration (USA)
Euratom: the European Atomic Energy Community
excursion: a rapid increase of reactor power above the set level of operation, either deliberate (for experimental purposes) or accidental
exposure limit: see *dose limit*

F

fabrication plant: a plant in which fuel for nuclear reactors is manufactured (see p 48)
fail safe: a principle of engineering design in which, if a device (particularly a safety device) fails, it does so in a manner that produces a safer rather than a more dangerous condition
fall-out: deposition of radioactivity from the atmosphere as a result of the explosion of a nuclear weapon or accidental release
fast breeder reactor: a misnomer for a fast reactor that also breeds fissile material; the neutrons are fast (see p 26) but the breeding process is always very slow (see p 183)
fast fission: fission induced by fast neutrons, as in a fast reactor
fast neutrons: unmoderated fission neutrons travelling with a speed close to that with which they were emitted during the fission process (see p 26)
fast reactor: a nuclear reactor, containing no *moderator*, in which most of the fissions are caused by *fast neutrons*
fast reactor fuel cycle: a nuclear fuel cycle in which recovered plutonium is burned in fast reactors (see p 55)
fast reactor launch cycle: a nuclear fuel cycle in which thermal reactors are used to produce the plutonium needed to establish a fast reactor power programme (see p 56)
fault tree: a diagram showing possible initiating events and sequences of successive failures that could lead to a serious accident
FBR: *F*ast *B*reeder *R*eactor
fertile material: a material such as uranium 238 or thorium 232 which can be transformed into *fissile material* by neutron irradiation (see p 35)
film badge: a small plaque containing a masked piece of photographic film worn by people working in areas in which they are liable to be exposed to ionizing radiation; the degree of darkening of the film provides a measure of the dose received
fissile material: a material such as uranium 235 or plutonium 239 which readily undergoes neutron-induced fission
fission: the splitting of a heavy nucleus such as uranium 235 into two roughly equal fission fragments, accompanied by the release of fast neutrons and energy (see p 19)
fission bomb: a nuclear weapon that produces an explosion by the uncontrolled fission of uranium or plutonium
fission fragment: a middle-mass nucleus (invariably highly radioactive) produced by the fission of a heavy nucleus (see p 21)
fission neutrons: fast neutrons emitted during the fission of heavy nuclei (see p 20)
fission products: the radioactive middle-mass elements that are produced as by-products of the fission process in a nuclear explosion or in the fuel elements of a nuclear reactor (see p 44)
flask: a heavily shielded container used to store or transport radioactive material, especially spent nuclear fuel (see pp 52 and 163)
fossil fuel: the collective name for fuels (such as coal, oil and natural gas) derived from once-living organisms
fuel assembly: a group of nuclear fuel elements forming a single unit for purposes of charging or unloading a reactor
fuel channel: a channel in a reactor core designed to hold one or more fuel assemblies or fuel elements
fuel charge: see *charge*

fuel cycle: the sequence of steps involved in supplying, using and reprocessing the fuel used in nuclear reactors (see Chapter 3)
fuel element, pin, rod: a unit of nuclear fuel used in a reactor; a component of a fuel assembly or sub-assembly
fuel inventory: the total amount of nuclear fuel invested in a reactor, a group of reactors or an entire fuel cycle
fusion: a reaction between two light nuclei resulting in the production of a nucleus heavier than either, usually accompanied by the release of energy (see p 22)
fusion bomb: a nuclear weapon, usually triggered by a *fission bomb*, in which the bulk of the explosive power is produced by thermonuclear fusion of deuterium, tritium or other light elements
fusion reactor: a reactor (still in the early stages of development) in which it is hoped to produce energy by controlled thermonuclear reactions (see p 43)

G

gamma decay, emission: a type of radioactive decay in which an unstable nucleus emits a *gamma ray* (see p 61)
gamma ray: a high-energy, short wavelength parcel of electromagnetic radiation emitted from a nucleus during *gamma decay* (see p 61)
gas centrifuge: a system used to separate the isotopes of an element such as uranium by using the high centrifugal forces in a rapidly rotating cylinder (see p 47)
gas-cooled reactor: a nuclear reactor that employs a gaseous coolant such as carbon dioxide or helium (eg the *Magnox*, *AGR* and *HTR* systems)
gaseous diffusion: a process used to separate the isotopes of an element such as uranium by making use of the fact that gas molecules of different masses diffuse at different rates through a porous membrane (see p 47)
generating board: an organization (in the UK) responsible for the generation and distribution of electricity within the area under its jurisdiction
generating capacity: the peak power output that a generating plant, power station or group of power stations can produce; measured in MW or GW
genetic effects: the long-term effects of radiation on future generations as opposed to immediate or delayed *somatic effects* on individuals (see p 69)
gigawatt (GW): one million *kilowatts*; a unit used to measure the generating capacity of a power station or group of power stations
gigawatt hour (GWh): one million *kilowatt hours* (units of electricity); a unit commonly used to measure the electrical output of a power station or group of power stations over a period of time
glassification: see *vitrification*
glove box: a sealed work chamber in which alpha-emitting materials such as plutonium can be safely handled using rubber gloves fixed to ports in the walls (see p 56)
graphite: a black crystalline form of carbon used as a moderator and/or reflector of neutrons in several reactor systems
gray (Gy): the unit of absorbed dose in the *SI System* measuring the energy deposited per unit mass; equal to 100 *rads*

H

half-life: the time taken for half the atoms in a sample of radioactive material to undergo decay (see p 59)
hard: of radiation, having a relatively high penetrating power – eg high-energy gamma rays
heat exchanger: the part of a nuclear power station where the coolant transfers its energy to another coolant circuit or to water to raise steam (see p 28)
heavy hydrogen: a common name for *deuterium*
heavy water: water in which most of the ordinary hydrogen atoms have been replaced by deuterium atoms; used as a moderator in nuclear reactors because of its low neutron-absorbing properties (see p 31)

helium: an inert gas used as a coolant in some reactor systems, eg the *HTR* system (see p 31)
hex: a colloquial name for *uranium hexafluoride*
highly active waste: nuclear waste with a very high level of activity
hot: jargon for highly radioactive
HTR: *H*igh *T*emperature *R*eactor (see p 29)
hydrogen bomb: the popular name for a *thermonuclear bomb*

I

IAEA: *I*nternational *A*tomic *E*nergy *A*gency (a United Nations-sponsored body, based in Vienna, set up in 1957 in order to promote the peaceful uses of atomic energy) (see p 184)
ICRP: *I*nternational *C*ommission for *R*adiological *P*rotection
incident: a term applied to an accident at a nuclear plant
induced radioactivity: artificial radioactivity produced by (for example) irradiation with neutrons
INFCE: *I*nternational *N*uclear *F*uel *C*ycle *E*valuation Exercise (see pp 44 and 179)
installed generating capacity: the peak generating capacity of all the power stations owned by a generating board, power company or country at a given time
intermediate heat exchanger: a heat exchanger in which heat energy is transferred from a primary to a secondary coolant circuit, as, for example, in a LMFBR (see p 35)
ion: an atom which has lost one or more of its circumnuclear electrons and has therefore become positively charged
ionizing radiation: radiation which removes circumnuclear electrons from atoms, eg alpha, beta and gamma radiation and X-rays (see p 62)
irradiated fuel: fuel that has been irradiated with neutrons in the core of the nuclear reactor
irradiation: exposure to radiation; eg to neutrons in the core of a nuclear reactor
isotopes: atoms of the same element having different atomic weights owing to differences in the numbers of neutrons in their nuclei (see p 17)

J

JET: *J*oint *E*uropean *T*orus (see p 40)
joule (J): the unit of energy in the *SI System* (see p 19)

K

kilowatt (kW): 1000 *watts*
kilowatt hour (kWh): the amount of energy provided when one *kilowatt* of power is supplied for one hour – the standard unit used to measure consumption of electricity

L

Lawson's criterion: in fusion research, the critical conditions needed to produce net power output (see p 39)
leaching: the slow dissolution of a substance from a solid containing it, eg fission products from vitrified nuclear waste (see pp 168 and 171)
lead: a metal used as shielding against ionizing radiation – especially gamma rays
lethal dose: a whole-body dose of ionizing radiation large enough to produce death in the individual exposed to it (see p 67)
light water: ordinary water containing virtually no heavy hydrogen in its molecules, as opposed to *heavy water*
liquid metal coolant: the coolant (usually liquid sodium) used in fast reactors with small cores because of its excellent heat transfer properties (see p 35)
liquid waste: radioactive waste in liquid form – usually an aqueous solution
LMFBR: *L*iquid *M*etal *F*ast *B*reeder *R*eactor

load factor: the load factor of a generating plant is the ratio (expressed as a percentage) of units actually generated over a period to units that could have been generated had the plant operated continuously at full power

load following: varying the output of a power station or reactor to match the changing requirements of an electricity distribution system

localized dose: a dose of radiation confined to a particular part of the body of the individual exposed to it

low active waste: nuclear waste that can usually be discharged into the environment in a controlled manner because of the small amounts of radioactive material that it contains (see p 54)

LWR: *L*ight *W*ater *R*eactor – a reactor that uses light water both as moderator and coolant, eg the *PWR* (see p 31) and *BWR* (see p 33) systems

M

magnetic bottle: a shaped magnetic field used to contain a plasma

magnetic confinement, containment: in fusion research, the use of shaped magnetic fields to confine a plasma (see p 39)

magnox: an alloy of magnesium used as a cladding material in natural uranium reactors of the *Magnox* type

Magnox reactor: the type of natural uranium reactor used in the first generation of British nuclear power stations (see p 28)

man-rem: a unit used to measure the total dose received by a given population over a given period: if one million people each receive an average of 0.01 rems, for example, then we say that the total dose received by the population is 10,000 man-rems

mass defect: the difference between the mass of a nucleus and the sum of the masses of its constituent nucleons; it is related to the *binding energy* of the nucleus (see p 18)

mass number: the total number of *nucleons* (*protons* and *neutrons*) in a nucleus (see p 17)

megawatt (MW): one million *watts*

meltdown: a type of nuclear accident in which the fuel becomes so overheated that it melts and collapses into the base of the reactor (see p 155)

merit order: the ranking order used to determine which power stations in a grid-linked electricity generation system will be used to meet a given load demand; the stations that are highest in the merit order (ie produce the cheapest electricity) are used first

microcurie (μCi): one millionth of a *curie* (see p 59)

millicurie (mCi): one thousandth of a *curie* (see p 59)

millirem: one thousandth of a *rem* (see p 66)

mixed oxide fuel: reactor fuel consisting of intimately mixed oxides of plutonium and uranium (which may be depleted) (see p 35)

moderator: the material used in a thermal reactor to slow the fission neutrons down to thermal energies (see p 27)

molecule: the smallest particle of a chemical compound that retains the chemical properties of the compound (see p 15)

MPBB: *m*aximum *p*ermissible *b*ody *b*urden (eg of a radioisotope)

mtce: *m*illion *t*onnes of *c*oal *e*quivalent (see p 71)

multiple containment: the principle of engineering design in which radioactive materials in a nuclear reactor or other plant are prevented from escaping into the environment by a series of containment systems, so that if one is breached, the next one comes into service (see p 156)

MW(e): megawatts (electrical) – a unit used to measure the electrical output of a nuclear reactor

MW(t): megawatts (thermal) – a unit used to measure the thermal (heat) output of a nuclear reactor

N

natural radioactivity: radioactivity in a naturally occurring as opposed to a man-made radioisotope

natural uranium: uranium having the isotopic proportions found in nature, ie 99.3 per cent U238 and 0.7 per cent U235

neutron: an elementary particle with mass number one and no electric charge – one of the constituents of the atomic nucleus (see p 17)

neutron bomb: a type of nuclear weapon that produces a comparatively low explosive yield but emits a very high number of potentially lethal neutrons

neutron moderator: see *moderator*

neutron reflector: see *reflector*

NII: *N*uclear *I*nstallations *I*nspectorate (UK) (see p 157)

Non-Proliferation Treaty: an international treaty, endorsed by the United Nations in 1968, designed to prevent the spread of nuclear weapons (see p 184)

NRC: *N*uclear *R*egulatory *C*ommission (USA)

NRPB: *N*ational *R*adiological *P*rotection *B*oard (UK)

nuclear binding energy: see *binding energy*

nuclear fission: see *fission*

nuclear force: the powerful force that binds the particles of the nucleus together (see p 17)

nuclear fuel cycle: see *fuel cycle*

nuclear fusion: see *fusion*

nuclear park: a (hypothetical) group of nuclear power stations, together with their fuel manufacturing and reprocessing facilities, located on a single site

nuclear pile: an outdated name for a *nuclear reactor*

nuclear reactor: a plant in which a neutron-induced fission chain reaction can be maintained, controlled and used (see p 26)

nuclear weapon: a fission or fusion bomb or warhead

nucleon: a proton or neutron, ie one of the particles that make up the nucleus of an atom (see p 17)

nucleus: see *atomic nucleus*

O

once-through fuel cycle: a type of fuel cycle in which spent fuel is stored in ponds rather than reprocessed (see p 45)

operational life: the period over which a plant such as a nuclear reactor is capable of performing its design function economically

oxide fuel: nuclear fuel manufactured from the oxide of a fissile material – generally in the form of a heat-resistant ceramic material (see p 29)

P

permitted dose: the dose of radiation that an individual (or population) is allowed to receive over a given time (see p 142)

PFR: *P*rototype *F*ast *R*eactor (see p 125)

pile: see *nuclear pile*

pinch effect: the process by which a plasma carrying an electric current is constricted due to the action of the magnetic field that the current produces

pitchblende: a particularly rich ore of uranium

plasma: an electrically neutral gas containing free positive ions and the electrons that have been removed from them (see p 38)

plutonium: a transuranic (ie artificial) element that readily undergoes neutron-induced fission (see p 35)

plutonium economy: a hypothetical situation in which a society depends for an important part of its energy supply on plutonium created and burned in fast reactors

pond: see *cooling pond*

positron: a positive electron – the anti-particle of the ordinary negative electron
pressure vessel: a reactor containment vessel, usually made of thick steel or pre-stressed concrete, capable of withstanding high internal pressures; used in gas-cooled and light water reactors
pressurizer: a component of the coolant circuit in a pressurized water reactor that is used to maintain the coolant pressure at the required value
primary energy source, fuel: a basic, naturally occurring energy source or fuel such as coal, oil, natural gas or uranium as opposed to a secondary (man-made) energy source or fuel such as electricity, coke or manufactured gas
proliferation: the acquisition of nuclear weapons by a state that did not previously possess them, whether by its own efforts, by purchase or gift or by illicit means
proton: an elementary particle with mass number one and unit positive electric charge – one of the constituents of the atomic nucleus (see p 17)
prototype reactor: a reactor built as a prototype for a larger, commercial reactor of the same basic type
proven reserves: reserves of a fuel or raw material that have actually been discovered and have had their extent appraised
PWR: *P*ressurized *W*ater *R*eactor (see p 31)

Q

quality factor: a factor used to express the relative effectiveness of different types of ionizing radiation in harming living tissue (see p 66)

R

rad: a unit of absorbed radiation dose, equivalent to 0.01 joules of absorbed energy per kg of irradiated material (see p 65); see also *gray*
radiation: in the nuclear context, high-energy electromagnetic waves such as X-rays and gamma rays or streams of fast-moving particles such as alpha or beta particles or neutrons (see p 62)
radiation area: an area to which access is controlled because of a local radiation hazard (see p 147)
radiation burn: skin damage akin to a heat burn produced by intense or prolonged exposure to ionizing radiation (see p 67)
radiation sickness: the acute effects caused by a large whole-body dose of ionizing radiation (see p 67)
radioactive: possessing or pertaining to radioactivity
radioactive area: see *radiation area*
radioactive decay: the process by which a nucleus moves to a more stable configuration by spontaneously emitting an alpha or beta particle or a gamma ray (see p 59)
radioactive source: a quantity of radioactive material intended for use as, or acting as, a source of ionizing radiation
radioactivity: see *radioactive decay*
radiography: a method of studying the interior of a specimen by passing penetrating radiation such as X-rays, γ rays or neutrons through it so that 'shadows' are cast by the internal structure
radioisotope: short for radioactive isotope
radiological protection: protection of nuclear workers etc from the harmful effects of exposure to ionizing radiation
radiological survey: a survey of the radiation present in a given area, eg in the vicinity of a nuclear plant, usually for *radiological protection* purposes
radiology: the branch of medicine that deals with the uses of ionizing radiation for medical diagnosis or treatment and the effects of such radiation on living materials
reactor: see *nuclear reactor*

reactor vessel: the container in which the fuel, moderator (if any), coolant and control rods of a nuclear reactor are situated

recoverable reserves: of a fuel such as uranium, the resources that can be recovered at a cost below a specified limit using existing technology

reflector: an extra layer of moderator or other material placed outside the actual core of a nuclear reactor in order to scatter back ('reflect') some of the neutrons that would otherwise escape

rem: *r*oentgen *e*quivalent *m*an: a unit of dose equivalent, being the product of the dose in *rads* and the *quality factor* (see p 66)

reprocessing: treating spent nuclear fuel in order to recover unburnt uranium and plutonium and isolate dangerous waste products (see Chapter 3)

reprocessing cycle: a nuclear fuel cycle in which spent fuel from thermal reactors is reprocessed without the plutonium thus recovered being actually used (see p 50)

reprocessing plant: a plant in which nuclear fuel is reprocessed (see p 52)

research reactor: a reactor built purely for research purposes rather than to produce usable power on a large scale

roentgen: a measure of the amount of ionization produced in a unit mass of air by radiation – the original unit used to measure radiation

S

self-sustaining: of a chain reaction, capable of maintaining itself

SGHWR: *S*team-*G*enerating *H*eavy *W*ater *R*eactor (see p 33)

shielding: a layer of material (eg lead or concrete) placed around a nuclear reactor, nuclear plant or source of radiation in order to contain the radiation

shut-down: stopping the chain reaction in a nuclear reactor by inserting the control rods or shut-down rods (see p 27)

shut-down rods: rods of strongly neutron-absorbing material that can be inserted into the core of a nuclear reactor in order to shut down the chain reaction (see p 27)

sievert (Sv): the SI unit of dose equivalent; equal to 100 *rems* (see p 65)

SI System: the internationally accepted system of units used in science

slow neutrons: another term for *thermal neutrons*

sodium: a metal with a low melting point used as a coolant for fast reactors (see p 35)

soft radiation: radiation having little penetrating power

solid waste: nuclear waste in solid form, eg the croppings from spent fuel cans

solvent extraction: a chemical separation process used in nuclear fuel reprocessing (see p 53)

somatic effects: the immediate and delayed effects of radiation on individuals as opposed to the long-term *genetic effects* on future generations

spent fuel: fuel that has been irradiated in a nuclear reactor to the required degree of burn-up

steam generator: the section of plant in a nuclear power station used to raise steam to drive the turbines

stockpile: to build up a store of a material such as plutonium or depleted uranium without actually using it

storage pond: a pond in which spent fuel elements are stored, eg in the once-through fuel cycle (see p 50)

sub-critical: containing insufficient fissile material or having insufficient reactivity to sustain a chain reaction

sub-critical mass: a mass of fissile material that is too small to sustain a chain reaction

T

thermal fission: fission produced by thermal (slow) neutrons rather than fast neutrons (see p 26)

thermal neutrons: fission neutrons that have been slowed down to thermal energies by means of a *moderator* (see p 26)

thermal output: the amount of heat energy produced by a nuclear reactor as opposed to its ultimate electrical output

thermal reactor: a nuclear reactor in which the fission chain reaction is largely maintained by *thermal* (slow) as opposed to *fast* neutrons

thermal recycle fuel cycle: a fuel cycle in which plutonium recovered from spent thermal reactor fuel is burned in thermal rather than fast reactors (see p 55)

thermal shield: a shield of thick metal plates placed inside the biological shield of a nuclear reactor in order to protect it from damage by overheating

thermonuclear fusion: fusion produced by heating a plasma up to a very high temperature (see p 38)

thermonuclear reaction: a nuclear reaction brought about by heating the components up to a very high temperature, so that the nuclei collide with high relative velocities (see p 38)

thermonuclear weapon: another name for a fusion bomb or warhead

thorium: a fertile material that can be converted into fissile U233 by neutron irradiation; a possible source of fuel for nuclear reactors

THORP: *th*ermal *o*xide *r*eprocessing *p*lant (see p 52)

tokamak: a toroidal magnetic containment system currently being investigated with a view to developing a fusion reactor (see p 39)

tonne: 1000 kilogrammes; sometimes referred to as a metric ton

toroid, torus: a doughnut-shaped body, chamber or magnetic field

transport flask: a massive, heavily shielded container used to transport spent nuclear fuel (see pp 52 and 163)

transuranic element: an artificial element with an atomic number greater than that of uranium (92), the heaviest naturally occurring element

trip: rapid automatic shut-down of a reactor brought about when one of its operational parameters (eg its temperature) deviates from its pre-set limit

tritium: the isotope of hydrogen having a mass number of three (see p 18)

U

UKAEA: *U*nited *K*ingdom *A*tomic *E*nergy *A*uthority

uncontrolled fission: a fission chain reaction which builds up rapidly in an uncontrolled manner as in a nuclear explosion

underground disposal: disposal of nuclear waste by permanent storage in an underground chamber (see p 55)

UNSCEAR: *U*nited *N*ations *S*cientific *C*ommittee on the *E*ffects of *A*tomic *R*adiation

uranium: an element normally consisting of 99.3 per cent fertile U238 and 0.7 per cent fissile U235 – the main current source of nuclear reactor fuel

uranium hexafluoride: a gaseous compound of uranium used as feedstock in uranium enrichment plants (see p 46)

USAEC: *U*nited *S*tates *A*tomic *E*nergy *C*ommission (replaced in 1975 by *ERDA* and *NRC*)

V

vitrification: a technique for the conversion of highly active liquid waste into glass currently being developed in Britain and France (see p 55)

W

waste: unwanted radioactive materials produced as by-products of the nuclear power industry (see p 54)

waste disposal: the disposal of radioactive nuclear waste under controlled conditions (see p 54)

waste management: see *waste disposal*

watt (W): the unit of power corresponding to an expenditure of one joule of energy per second

whole-body dose: a dose of radiation to which the whole body of the individual concerned is exposed

X

X-rays: very short wavelength electromagnetic radiation originating from the circumnuclear electrons of atoms rather than from within their nuclei, as in the case of gamma rays

Y

yellow cake: concentrated crude uranium oxide – the form in which most uranium is shipped from mining areas to fuel fabrication plants

Z

zircaloy: an alloy of zirconium and aluminium used as fuel cladding in water reactors such as the PWR system

Bibliography

In order to help any readers who would like further information on the topics covered by our book, we have prepared the following bibliography. This is written in two sections: Section A lists recently published books and booklets on nuclear power and energy, and Section B, sources of further information relating to specific chapters. Both lists represent a wide spectrum of ideas and, together, cover all the various aspects of the nuclear debate.

Section A: Recent books and booklets on nuclear power and energy

Pro-nuclear

Beckmann, Peter (1979) *The Non-problem of Nuclear Wastes.* Boulder, Colorado: Golem Press

Greenhalgh, G (1980) *The Necessity for Nuclear Power.* London: Graham & Trotman Ltd

Hoyle, Fred (1977) *Energy or Extinction: The Case for Nuclear Energy.* London: Heinemann

Hoyle, Fred and Hoyle, Geoffrey (1980) *Commonsense in Nuclear Energy.* London: Heinemann

Longstaff, Michael (1980) *Unlocking the Atom – A Hundred Years of Nuclear Energy.* London: Frederick Muller Ltd

UKAEA (1981) *Nuclear Power – The Real Facts.* London: UKAEA

Wyatt, Alan (1978) *The Nuclear Challenge – Understanding the Debate.* Toronto: Book Press Ltd

Anti-nuclear

Bunyard, Peter and Morgan-Granville, Gerald (1980) *Nuclear Power – What It Means to You.* Ecoropa

Garrison, Jim (1980) *From Hiroshima to Harrisburg.* London: SCM Press Ltd

Patterson, Walter C (1976) *Nuclear Power.* Harmondsworth: Penguin

Students Against Nuclear Energy (1980) *Anti-Nuclear Now . . . or Never.* London

Balanced

Foley, Gerald and van Buren, Ariane (eds) (1978) *Nuclear or Not? Choices for our Energy Future.* London: Heinemann

Titterton, Ernest and Roebotham, F P J (1979) *Uranium: Energy Source of the Future.* The Australian Institute of International Affairs

Williams, Roger (1980) *The Nuclear Power Decisions.* London: Croom Helm Ltd

General

Chapman, Peter (1975) *Fuel's Paradise – Energy Options for Britain.* Harmondsworth: Penguin

Crabbe, David and McBride, Richard (1978) *The World Energy Book.* London: Kogan Page

Curran, Samuel C and Curran, John S (1979) *Energy and Human Needs.* Edinburgh: Scottish Academic Press

Foley, Gerald (1976) *The Energy Question.* Harmondsworth: Penguin

Goodman, G T and Rowe, W D (eds) (1979) *Energy Risk Management.* London: Academic Press Inc

Ruedisili, Lon C and Firebaugh, Morris W (eds) (1978) *Perspectives on Energy – Issues, Ideas and Environmental Dilemmas.* Oxford: Oxford University Press

World Energy Conference (1979) *World Energy: Looking Ahead to 2000.* Guildford and New York: IPC Science and Technological Press

Section B: Sources of further information relating to specific chapters

Chapters 1 and 2

Nero, Antony V Jr (1979) *A Guidebook to Nuclear Reactors.* University of California Press

UKAEA (1980) *Nuclear Energy: Today's Power for Tomorrow's Generation.* London: UKAEA

UKAEA (1977) *Nuclear Power Reactors.* London: UKAEA

Energy Research and Development Administration (1975) *Advanced Nuclear Reactors – An Introduction.* Washington: ERDA

UKAEA (1975) *Fast Reactors in Britain.* London: UKAEA

Chapter 3

IAEA (1980) *INFCE Reports.* Vienna: IAEA

Marshall, Walter (1980) The use of plutonium. *Atom*, No 282 (April)

Chapter 4

IAEA (1979) *Radiation – A Fact of Life.* Vienna: IAEA

IAEA (1981) *Facts about Low-level Radiation.* Vienna: IAEA

NRPB (1981) *Living with Radiation.* NRPB

Chapter 5

HMSO (1980) *UK Digest of Energy Statistics.* London: HMSO

UN (1970-79) *UN Statistical Year Books.* UN

BP (1979) *BP Statistical Review of the World Oil Industry.* London: British Petroleum Co Ltd

Energy Information Administration (1978) *Annual Report to Congress: Statistics and Trends of Energy Supply, Demand and Prices.* May

Chapter 6

Gerald Leach *et al* (1979) A low energy strategy for the United Kingdom. *IIED Science Reviews*

BP (1979) *Oil Crisis . . . Again?* London: British Petroleum Co Ltd

Final Report by Energy Project of Ford Foundation (1974) *A Time to Choose: America's Energy Future.* Cambridge, Mass: Ballinger

Chapter 7

HMSO (1978) *The Development of Alternative Sources of Energy,* Cmnd 7236. London: HMSO

Clarke, F J P (1977) Status report on the alternative energy resources. *Atom*, No 252 (October)

Leach, Gerald (1981) Energy futures. *Atom*, No 294 (April)

Chapter 8

Jensen, W G (1969) *Nuclear Power.* Henley-on-Thames: G T Foulis and Co Ltd

Chapter 9

OECD (1979) *Uranium Resources, Production and Demand.* Paris: OECD

CEGB *Annual Report 1979-80* (Appendix 3)

Chapters 10 and 11

United States Nuclear Regulatory Commission (1975) *Reactor Safety Study: An Assessment of Accident Risks in US Commercial Nuclear Power Plants.* WASH 1400

IAEA (1979) *Radiation – A Fact of Life.* Vienna: IAEA

IAEA (1981) *Facts about Low Level Radiation.* Vienna: IAEA

Gloag, Daphne (1980) Risks of low level radiation – the evidence of epidemiology. *British Medical Journal,* **281**, (November)

Gloag, Daphne (1980) Radiation exposure and the protection of the community. *British Medical Journal,* **281**, (December)

Commission of the European Communities (1980) *Nuclear and Non-nuclear Risk – An Exercise in Comparability.* EUR 6417 EN 1980

Fryer and Griffiths *World Wide Data on the Incidence of Multiple Fatality Accidents,* SRD R149. London: HMSO

Fremlin, J H (1980) Risk relativities. *Atom,* No 283 (May)

Roberts, L (1979) Radioactive waste – policy and perspective. *Atom,* No 267 (January)

Chapter 12

Foley, Gerald and van Buren, Ariane (eds) (1978) *Nuclear or Not? Choices for our Energy Future* London: Heinemann

SIPRI (1979) *Nuclear Energy and Nuclear Weapon Proliferation.* London: Taylor & Francis

Marshall, W (1978) Proliferation and the recycling of plutonium. *Atom*, No 263 (September)

IAEA (1980) *INFCE Reports.* Vienna: IAEA

Index